Let Them Eat Precaution

How Politics Is Undermining the Genetic
Revolution in Agriculture

Edited by
Jon Entine

The AEI Press

Publisher for the American Enterprise Institute
WASHINGTON, D.C.

Distributed to the Trade by National Book Network, 15200 NBN Way, Blue Ridge Summit, PA 17214. To order call toll free 1-800-462-6420 or 1-717-794-3800. For all other inquiries please contact the AEI Press, 1150 Seventeenth Street, NW, Washington, DC 20036 or call 1-800-862-5801.

This publication, and the research and conference that led to it, were funded by the American Enterprise Institute's National Research Initiative and by AEI's Inez and William Mabie Endowment for Agricultural Policy Research.

Library of Congress Cataloging-in-Publication Data
    Let them eat precaution: how politics is undermining the genetic revolution in agriculture / edited by Jon Entine
        p. cm.
    Includes bibliographical references and index.
    ISBN 0-8447-4200-7 (cloth: alk. paper)
    1. Plant biotechnology. 2. Plant biotechnology—Social aspects.
   3. Transgenic plants. 4. Transgenic plants—Risk assessment. 5. Food supply.
   I. Entine, Jon.

SB106.B56L48 2005
631.5'233—dc22

2005007449

11 10 09 08 07 06   1 2 3 4 5 6

Printed in the United States of America

Contents

Editor's Note

Let Them Eat Precaution grew out of a recent conference on "Food Biotechnology, the Media, and Public Policy," held at the American Enterprise Institute. It has become clear in recent years that the most immediate impact of the genetic revolution is being experienced in agricultural and food production. Genetic technology has resulted in the development of more nutritious food at far less environmental cost. Yet the gene revolution in food is mired in controversy.

The AEI forum, designed to address this ideological impasse, drew together a uniquely diverse array of scientists, consumer activists, academicians, government policymakers, specialists in public opinion and public relations, and representatives from affected industries. Each chapter of *Let Them Eat Precaution* is based on a paper originally presented at the conference, but includes recent developments in the connection between genes, agriculture, food, and nutrition. These essays offer an essential resource for policymakers, journalists, and interested citizens concerned about food, agricultural production, and genetic policy.

The book is divided into three sections. "Ideological Gridlock" explores the political and cultural reasons behind the current standoff between scientists and some consumer groups. "Consequences" examines the impact of this gridlock on society, particularly in the developing world, with a new generation of bioengineered products on the horizon. In "Solutions," we review provocative avenues out of this political impasse.

I would like to thank Robert Paarlberg for graciously allowing AEI to borrow the apt title of his article for this book. *Let Them Eat Precaution* captures the dilemma faced by millions of people in the developing world who are being blocked from the opportunity of sharing in the bounty offered by the genetic modification of crops and foods.

JON ENTINE
Adjunct Fellow, American Enterprise Institute
Scholar in Residence, Miami University (Ohio)

Introduction:
Beyond Precaution

Jon Entine

The debate over food and genes has degenerated in recent years into a cartoon discourse. Common sense, science, and the needs of the poor and malnourished are now regularly sacrificed to political calculations.

Consider the contretemps over "Golden Rice," the genetically modified, vitamin-enhanced version of the world's most popular staple. White rice makes up 72 percent of the diet of the people of Bangladesh, and nearly as much in Laos and Indonesia; more than 40 percent in Madagascar and Sierra Leone; and around 40 percent in Guyana and Suriname. While it is a filling food that can be grown in abundance, it has a major drawback: It lacks vitamin A. Vitamin A deficiency (VAD) weakens the immune system, increasing the risk of infections such as measles and malaria. Severe deficiencies can lead to blindness. Children and pregnant women are particularly vulnerable to VAD. According to the World Health Organization, there are more than 100 million VAD children around the world. Some 250,000 to 500,000 of these children become blind every year; half of them die. In Asia and Africa, nearly 600,000 women with vitamin A deficiencies die from childbirth-related causes (World Health Organization 2004).

How should we as a society respond to a crisis of such malignant proportions?

In the 1960s, the developed world began reaping enormous benefits from the "Green Revolution," with the widespread use of pesticides, fertilizers, sophisticated irrigation, mechanization, and the use of new crop cultivars that dramatically improved yields and the nutritional content of crops. Norman Borlaug won the Nobel Peace Prize for his development of high-yielding wheat varieties that helped feed people in developing countries. Now, with the advent of agricultural biotechnology, we have an opportunity to extend those gains with the development of new grains, fruits, vegetables, and other nutritionally enhanced foods developed using far fewer pesticides.

The engineering of so-called Golden Rice stands as the most powerful example of the potential of this remarkable technology. In 1999, Swiss and German scientists developed this unique gold-colored rice, the first major genetically enhanced food. Until this point, bioengineers had focused on developing new strains of crops and grains, such as cotton, wheat, and soybeans; this was the first time this technology was used to develop a food that people eat directly, rather than a crop that needed to be processed. The new variety was produced by splicing into white rice two genes from the daffodil, which give the rice a golden color, and one from a bacterium. The added genes cause the new rice strain to produce beta carotene, which the body can convert to vitamin A. Newer varieties have been created to add iron and to make the iron already in the white rice more readily absorbed into the body. Yet, despite the enormous promise of Golden Rice and other remarkable new crops and foods, the biotech phase of the Green Revolution became mired in controversy.

The central opposition comes from organized antibiotechnology activists. They include well-known environmental groups such as Greenpeace, Friends of the Earth, and organic advocates; religious groups such as Christian Aid; and a small but media-savvy sector of the investment community known as "socially responsible investors," which includes groups such as Co-Op America; the Interfaith Center on Corporate Responsibility (ICCR), a religious-based advocacy group; and the Social Investment Forum, the trade group for liberal social investors. This loosely organized coalition attacks vitamin-packed rice seed and other biotech products on two grounds.

First, they contend that genetic technology is inherently unpredictable, conjuring up images of a genetic Godzilla that could cause irreparable environmental and health damage. They argue that genes not subject to checks and balances in nature could be released into the environment, causing untold havoc, although scientists believe this scenario is unlikely outside of doomsday scenarios that could apply just as well to conventional agriculture.

And as the authors of the chapters in this volume make clear, agricultural biotechnology has proven to be even less risky than conventional gene transfer techniques developed before the advent of genetic modification, including radiation and selective breeding, to produce new varieties with enhanced nutritional qualities or disease resistance.

Second, they argue that biotechnology crops and foods will ultimately result in higher food prices and less variety because private companies patent many of the new seeds and food genes. Some have even gone so far as to dismiss Golden Rice as bad science and a "gift horse for the poor" (Institute of Science in Society 2005). Other critics imply that Golden Rice amounts to a "Trojan Horse"—as Genewatch, one British-based antibiotech group, characterizes it—designed to soften opposition to agricultural biotech (Fumento 2003).

"At the end of the day," concludes a briefing on Grain.org, a popular antibiotechnology Web site, "the main agenda for golden rice is not malnutrition but garnering greater support and acceptance for genetic engineering amongst the public, the scientific community and funding agencies" (Grain.org 2001).

In fact, Golden Rice was developed over ten years at a total cost of $2.6 million, using funds donated by the Rockefeller Foundation, the Swiss Federation, the National Science Foundation, and the European Union. Numerous corporations have contributed to its development, donating their expertise in what is called "open source" drug discovery, modeled after the open source technology that was used in computing to develop the Linux operating system (Pollack 2005). In May 2004, the biochemical firm AstraZeneca announced it would be distributing one type of genetically modified rice, developed by two European scientists, to Asian countries free of charge because of the crop's health benefits. Three months later, biotech giant Monsanto announced it would be providing royalty-free licenses for all of its patented technologies that might help further the development of this particular crop. Still, because of the intense opposition to biotechnology, it may take four years or more before the new crop can wind its way through the Byzantine regulatory system and go into production.

Increasingly, nongovernmental organizations (NGOs) and religious groups that often align themselves on other issues with Greenpeace and other more radical environmental groups have come out in favor of agricultural biotechnology.

"I think [Golden Rice] has tremendous potential," said Alan McHughen, a senior research scientist at the University of Saskatchewan in Canada and author of *Pandora's Picnic Basket: The Potential and Hazards of Genetically Modified Foods*. "Many expert ethicists, including the Nuffield Foundation, the Church of England and even the Vatican have given their seal of approval for genetically modified food that is used to provide more food or more nutritious food" (Orfinger 2004).

Agricultural Biotechnology on Hold

The worldwide $46.6 million commercial biotechnology industry rests on four crops: cotton, used to make clothing, and soybeans, corn, and canola, used primarily for animal feed or to make oil and other ingredients for processed food. These four products increase yields by programming plants for two new qualities: to generate "natural" insecticides to ward off killer pests, or to be herbicide resistant, so farmers can spray insecticides without killing the plants. These innovations reduce the overall need for costly and potentially harmful chemicals, a dramatic environmental achievement in its own right.

Although the first generation of biotech crops increased yields and reduced the use of costly pesticides, they came under heavy criticism from antibiotech activists because they were developed by private corporations and used mostly by corporate farmers. The new round of innovations, however, is specifically designed to target malnutrition and the needs of poor farmers, and to attack broader environmental concerns. These new products include drugs, known as functional foods and nutraceuticals, made from genetically modified plants, as well as crops that are more pest and drought resistant, nutritionally enhanced crops and foods, and even forest trees that are being tweaked to extract toxins from the soil, resist disease, and absorb carbon to help reduce global warming.

None of these innovations has yet appeased antibiotech protestors. The controversy is frozen over the issue of whether the process of producing the products or the products themselves might result in unacceptable environmental or health hazards. Critics and defenders cannot even agree on whether this new technology is evolutionary or revolutionary. Farmers and plant breeders have relied for centuries on crossbreeding, hybridization, and other forms of genetic modification to improve the yield, quality, and disease resistance of crops. Today, virtually every plant grown commercially for food or fiber is a product of crossbreeding, hybridization, or both. Many of the same techniques have been used in modifying animal breeds and developing new pharmaceuticals.

Most geneticists and government regulators believe agricultural biotechnology is benign and at least as safe as traditional breeding techniques. There is absolutely no evidence that genetic modification poses greater risks than crossbreeding and gene-splicing, which have given us such products as the tangelo and seedless grapes. Using such traditional methods (including the radioactive bombardment of plants to create mutations, a scattershot process embraced by antibiotech groups as a "safe" alternative to biotechnology), thousands of genes, often of unknown function, are moved into crops. The new tools of biotechnology allow breeders to more precisely select single genes that produce desired traits and move them from one plant or animal to another. Researchers have been able to supercharge an organism's natural defenses using genetic material already in place or by introducing genes from other plants or animals.

Even the decision in May 2004 by the United Nations food agency, the Food and Agriculture Organization, endorsing the safety and health benefits of biotechnology in crops and food and urging its extension to the developing world, has not allayed the concerns of protesting NGOs. Consequently, while agricultural biotechnology continues to spread, it remains limited to but a select few countries: the United States, Argentina, Canada, Brazil, China, and South Africa. Almost all of the acreage remains devoted to only the four first-generation biotech crops: cotton, soybeans, corn, and canola.

There are serious issues about biotechnology and public policy that deserve vigorous public discussion. Legitimate questions have been raised about the degree to which corporations should be allowed to patent and therefore control beneficial biotech seeds or products they develop. Patents on human genes are the new stock in trade for biotechnology companies. The number of agriculture-related patents soared from next to nothing in 1986 to more than seven hundred annually by the year 2000.

While genetically modified (GM) crops offer many benefits, farmers who utilize these seeds can do so only after agreeing to certain licensing agreements. Universities, the federal government, and various public entities control many of the key technologies necessary for agricultural biotech (approximately 24 percent), but private companies own three-quarters of U.S. patents, including the most relevant genes. Monsanto (which controls 14 percent), Dupont (13 percent), Syngenta (7 percent), Bayer (4 percent), Dow (3 percent), and myriad other firms take the position that they need to recoup their research costs (Hayden 2004).

Unfortunately, reasonable concerns over patent rights, which should be part of the public debate, have been obscured by sensational and often misleading allegations. Critics claim that while the process of genetic modification might appear to be more precise than conventional breeding techniques, it is dangerous because of what they say is a limited understanding of potential hidden problems from allergens or antibiotic resistance. They often invoke the so-called precautionary principle to sow doubt about the long-range impact of this technology.

The Alice-in-Wonderland World of the "Precautionary Principle"

The precautionary principle, first used as a legal principle in Sweden in the late 1960s and Germany in the 1970s during environmental protests, has gradually become the weapon of choice for groups dedicated to scuttling technology they believe entails some risk—any risk. Although its adherents claim it asserts nothing more revolutionary than the maxim "Better safe than sorry," this innocent ring of moderation is deceptive in practice.

Precaution can be a prescription for paralysis. After all, innovation by definition entails unknown risks. Rather than encouraging a calm assessment of a complicated technology, the precautionary principle ends up firing hysteria. It exploits the public's inability to balance health and environmental benefits against a reasonable assessment of environmental risk. It also does not take into account the potential human cost of not innovating. That's one reason the World Trade Organization (WTO) forbids any nation from banning imports unless it can be proved to a "scientific certainty" that products are unsafe (World Trade Organization 1999).

Slavish adherence to an ultraconservative precautionary principle merely shifts risk to some current practices that would be banned if the principle were actively enforced. While there have been no documented health problems, deaths, or injuries linked to bioengineering, dozens of people die every year from eating organic and "natural" products contaminated as the result of poor quality control. Recall the dozens of serious injuries and the death of a Seattle girl in 1997 from drinking unpasteurized juice contaminated with E. coli, made by the Odwalla company from apples that had fallen in bacteria-laden "natural" fertilizer—that is, dung (Entine 1999). If the precautionary principle were applied to "natural" foods, they would be stripped from the grocery shelves overnight.

Claiming to be acting on behalf of innocent but unaware consumers and the "natural environment," determined protestors attempt to co-opt the debate by invoking the precautionary principle and using incendiary pejoratives like "pollution" and "contamination" to describe anything that contains genetically modified seeds or ingredients. That's exactly what has happened in an ongoing protest against farmers using a bioengineered treatment to increase milk yields. More than a decade ago, farmers discovered that cows given recombinant bovine somatotropin—rBST—produce more milk for a longer time. That means less feed and fuel are used compared with lower-producing herds, resulting in a host of environmental benefits. But the biofermentation production process, which is similar to making beer and wine and doesn't change the milk, involves biotechnology; and so organic and antibiotech activists allege that 90 percent of our milk supply is hopelessly "contaminated" by being mixed with milk from cows treated with a protein supplement.

Are milk drinkers endangered? Not according to independent studies in the United States and Europe. Consumer Reports, which summarized the scientific consensus, concluded "milk from hormone-treated cows poses no appreciable risk to humans" (Consumer Reports 2000).

Time and again, dire antibiotech warnings have proved feckless. In one well-known episode in 1999, an international firestorm flared when a letter was published in the magazine Nature suggesting that the monarch butterfly might face some danger from exposure to Bt corn pollen. This was not a peer-reviewed article or a state-of-the-art study, but a short summary of a four-day laboratory test. Its author, John E. Losey, carefully wrote: "It would be inappropriate to draw any conclusions about the risk to monarch populations in the field based solely on these initial results" (Losey, Rayor, and Carter 1999). In the hands of those claiming to be environmentalists, these modest concerns mutated into near hysteria. Activists outfitted with wings were dispatched to protests to die on cue to illustrate the fate facing humanity if scientists were not restrained and our dalliance with agricultural biotechnology not abandoned.

Backed into a corner by public concern, the National Academy of Sciences launched a two-year study of the monarch butterfly "crisis" (Sears et al. 2001). The detailed—and costly—report concluded that "the portion of the monarch population that is potentially exposed to toxic levels of Bt corn pollen is negligible" and therefore the risks to the butterfly from GM corn "should remain very low." But while the antibiotech smoking gun proved to have no bullets in this and similar cases, the damage to public discourse was already done. Years of exaggeration and misinformation have taken an enormous toll, undermining public confidence in science and genetics, profoundly altering the trajectory of biotechnology applications, and damaging the financial wherewithal of dozens of companies and university research projects.

The potentially devastating effect of the precautionary principle can be seen in the Philippines, where 42 percent of the diet comes from white rice. Scientists and UN food experts estimate that a broad acceptance of Golden Rice could avert 879 deaths, 1,925 corneal ulcers, and 15,398 cases of night blindness each year (Hayden 2004). Yet a Philippine-based antibiotechnology group, Masipag, with ties to Greenpeace and other international antibiotech campaigners, has aggressively lobbied against Golden Rice on the grounds that the benefits from beta-carotene are minimal and that Philippine farmers do not want to grow genetically modified crops.

Belying claims by the rice's ideological opponents, a survey of Philippine farmers reported in *Nature Biotechnology* found that most are not opposed to Golden Rice. "There is a huge disparity from what the anti-GMO groups are saying and what the farmers really have said in my research," says Mark Chong of Cornell University. "Most of the farmers know next to nothing about agricultural technology" (Chong 2003).

Farmers are mainly concerned with producing enough rice to meet immediate needs and would welcome the enhanced rice, says Chong. It is significant, he notes, that not a single barrio leader mentioned antibiotech nongovernmental organizations as a trusted information source, even though

Masipag operates programs in the heart of the rice-growing region. This casts doubt on the legitimacy of claims by antibiotech groups that they represent the broad concerns of Philippine farmers (Chong 2003). How farmers will react years from now when these new grains actually are cleared for planting remains an open question.

As is so common in this debate, the most heated claims by those challenging the science are often the ones least supported by scientists. "Greenpeace has a strategy to convince people that Golden Rice provides so little beta-carotene that it is useless," notes Ingo Potrykus, the Zurich-based researcher who helped developed the rice. "This group and its allies base their argument on 100 percent of the recommended daily allowance [RDA], thus hiding the fact that far lower values are effective against mortality, morbidity, and blindness. The Golden Rice . . . provide[s] true benefits at just 300 grams [10.5 ounces] per day" (Potrykus 2001).

Enviromanticism

In case after case, activist groups have demonized biotechnology by exploiting a general wariness about science. The 2004 National Science Foundation study of science and engineering indicators shows that, although Americans express strong support for science in the abstract, public knowledge about science issues and the process of science remains low and the public is increasingly turning to the Internet as a major source of information (National Science Foundation). Echoing the views of other antibiotech crusaders, Amory Lovins, founder of the Rocky Mountain Institute, waxes about the dangers of "replacing nature's wisdom with people's cleverness" (Lovins and Lovins 1999). For many biotech critics, this is not a scientific dispute but an ideological and religious one, driven by a simple—and dangerously simplistic—principle: Don't tamper with nature. It is a romantic and superficially seductive message, but a blanket insinuation that nature's products are always benign or better is obviously nonsense.

Some mainstream environmental groups, such as the Sierra Club, and social investors, who could have taken the high road on a complex issue, instead stand with antiscience hardliners. They often portray themselves as advocates of consumer choice by arguing for mandatory labeling of products made with genetically modified ingredients. At first blush, more disclosure seems reasonable and moderate. But will it provide any tangible benefit to consumers? As recently as July 2004, an independent panel of the National Academy of Sciences concluded without equivocation that genetically engineered crops do not pose any health risks that are not also present in conventionally produced crops.

"The most important message from this report is that 'It's the product that matters, not the system you are using to produce it,'" said Jennifer Hillard, a consumer advocate from Canada and one of the co-writers of the report *Safety of Genetically Engineered Foods* (Pollack 2004).

The labeling argument is a disingenuous ploy, as even its proponents acknowledge. A spokesperson for the Interfaith Center on Corporate Responsibility told me that mandatory labeling would be akin to slapping "a skull and crossbones" on GM products. And Michael Passoff of As You Sow, another antibiotech group, predicted to me that if the mandatory labeling campaign succeeds, "We expect that [the food industry] won't want to risk alienating their customers with labeling, so they'll eventually decide not to use any bio-stuff at all" (Entine 2002). In other words, GM products with absolutely no evidence of posing any danger, but with proven health and environmental benefits, would be vaporized from the marketplace.

While not a panacea, GM technology offers unique tools to address international food needs, especially in countries with increasing populations and widespread poverty. There are certainly valid concerns that need to be addressed if genetic modification is to get a fair shot in the marketplace. However, in the current atmosphere, rational policy initiatives and coordinated international trade policies are extremely difficult to undertake.

The unfulfilled potential of biotechnology might well rest on how it comes to be perceived by the greater world community. Public perceptions about bioengineering have dogged the research and commercialization strategies of the biotech industry since the first commercial products were introduced more than a decade ago. The often-politicized process has prompted firms to formulate what are known as "freedom to operate" strategies, which allow otherwise competing companies to cooperate in research without infringing on the patent rights of their research partners. In response, an international advocacy industry has coalesced, seeking to limit this freedom to operate in the name of social, environmental, and health responsibility.

This coalition includes traditional activists, such as public interest research groups, self-defined environmentalists, religious groups, social investment organizations, and umbrella antibiotech groups, like the GE Food Alert Coalition. They seek to apply public relations and, by proxy, financial pressures to influence the debate and public policy; many are determined to scale back radically or even kill the introduction of bioengineered products and processes. Also highly involved are the media, who have acted as a filter and sometime mouthpiece and advocate for the antibiotech perspective.

What appears to be lacking in the public debate in Europe and, increasingly, in the United States is a candid discussion about the current and potential benefits that these technologies can provide. That's where *Let Them Eat Precaution* has a role to play. The chapters in this book deconstruct the politics of the biotechnology debate and examine the extremely well-funded antibiotech industry; they also renew the promise of GM technology. Largely segregated to industrial crops in the developed world, there is now hope that the next generation of products—foods and crops that enhance nutrition or help in the development of critical new drugs—will break the public perception gridlock.

References

Chong, Mark. 2003. Acceptance of Golden Rice in the Philippine "Rice Bowl." *Nature Biotechnology* 21 (September 1): 971–72.

Consumer Reports. 2000. Milk Report: New Questions for an Old Staple. *Consumer Reports* 65, no. 1 (January): 34.

Entine, Jon. 1999. The Odwalla Affair: Reassessing Corporate Social Responsibility. *At Work* 8, no. 1 (January/February): 1–6. http://www.jonentine.com/articles/odwalla.htm (accessed February 13, 2005).

———. 2002. Dairy Report Milking the Public's Food Fears. *San Francisco Chronicle.* February 24.

Fumento, Michael. 2003. Plants That Will Save Lives and Eyes. *American Outlook Today.* http://www.hudson.org/index.cfm?fuseaction=publication_details&id=3094 (accessed February 13, 2005).

Grain.org. 2001. Grains of Delusion. Grain.org briefing. February 2001. http://www.grain.org/briefings/?id=18 (accessed February 13, 2005).

Hayden, Thomas. 2004. Seeds of Change. *Wired.* June, 152–53.

Institute of Science in Society. 2005. The "Golden Rice"—An Exercise in How Not to Do Science. http://www.i-sis.org.uk/rice.php (accessed February 13, 2005).

Losey, J. E., L. S. Rayor, and M. E. Carter. 1999. Transgenic Pollen Harms Monarch Larvae. *Nature* 399:214.

Lovins, Amory B., and L. Hunter Lovins. 1999. Replacing Nature's Wisdom with Human Cleverness. August 1. http://www.amberwaves.org/web_articles/lovins.html (accessed February 13, 2005).

National Academy of Sciences. 2004. *Safety of Genetically Engineered Foods: Approaches to Assessing Unintended Health Effects.* Washington, D.C.: National Academy of Sciences Press. http://www.nap.edu/openbook/0309092094/html (accessed February 13, 2005).

National Science Foundation. National Science Board. 2004. Science and Engineering Indicators 2004. Released May 4, 2004. http://www.nsf.gov/sbe/srs/seind04/start.htm (accessed February 13, 2005).

Orfinger, Becky. 2004. DisasterRelief.org. October 8. http://www.disasterrelief.org/Disasters/000808gmofoods/ (accessed February 13, 2005).

Pollack, Andrew. 2004. Panel Sees No Unique Risk from Genetic Engineering. *New York Times.* July 28.

———. 2005. Open-Source Practices for Biotechnology. *New York Times.* February 10.

Potrykus, Ingo. 2001. Interview by Michael Fumento. Golden Rice—A Golden Chance for the Undeveloped World. *American Outlook.* July–August. http://www.fumento.com/goldenrice.html (accessed February 13, 2005).

Sears, Mark K., et al. 2001. Impact of Bt Corn Pollen on Monarch Butterfly Populations: A Risk Assessment. *Proceedings of the National Academy of Sciences (PNAS)* 98, no. 21 (October 9): 11937–42.

World Health Organization. 2004. World Health Report. http://www.who.int/whr/2004/en/ (accessed February 12, 2005); see also Combating Vitamin A Deficiency. http://www.who.int/whr/2004/en/ (accessed February 12, 2005).

World Trade Organization. 1999. International Institute for Sustainable Development Report on the WTO's High-Level Symposium on Trade and Development. http://wto-org/hlms/sumhldev.htm (accessed February 13, 2005).

PART I

Ideological Gridlock

Introduction

For all its vast demonstrated value, agricultural biotechnology remains dramatically underutilized, mired in controversy around the world. The commentators in this opening section of *Let Them Eat Precaution* are Thomas Hoban, professor of food science and sociology and anthropology at North Carolina State University; Channapatna S. Prakash, professor of plant biotechnology at Tuskegee University, and Gregory Conko, director of food safety policy at the Competitive Enterprise Institute; and Tony Gilland, the science and society director of the British Institute of Ideas in London. They outline the current state of the technology and explain why the debate has played out so differently around the world.

While Europe has aggressively blocked the use of agricultural biotechnology, consumers in the United States have been somewhat indifferent, at least when GM crops are used as ingredients. More than 86 percent of soy, a key ingredient in thousands of food products, and 40 percent of corn are genetically modified. As a result, the federal government has been cautious about introducing what the biotech industry sees as intrusive oversight. In May 1992, the U.S. Food and Drug Administration (FDA) stated that it was not aware of any information showing that foods derived by these new methods differ from other foods in any meaningful or uniform way. The government claims that no substantive scientific evidence has been presented to justify altering that policy or requiring specific labeling of foods with GM ingredients.

In Britain and Europe, however, food scares during the 1990s undermined public confidence in government oversight and sowed confusion over the benefits and potential dangers of GM products, throwing a wrench into the regulatory machinery. The European Union has been observing an unofficial moratorium on new bioengineered food since 1998. Only in 2004 did the EU approve the importation of two genetically modified corn varieties made by Monsanto, based in the United States, and the Swiss biotechnology company Syngenta, although neither corn was approved for cultivation. The embargo on food with genetically modified ingredients remains in place and will not likely be lifted until the EU receives assurances that the United States won't resist its new strict labeling rules.

Japan, Korea, Australia, New Zealand, and other countries require, or have announced plans to require, labeling of GM-derived foods. A number of developing countries also support more restrictions on GM foods, mostly because they fear they could lose access to primary export markets in Europe if they should move to GM crops and anger EU countries. There is also a general unease in segments of the developing world, fed by European non-governmental organizations, that large corporations will develop too much power and control over seed technologies.

These varying perspectives are captured in "Global Views on Agricultural Biotechnology," by Thomas Hoban. The premier researcher in attitudes toward agricultural biotechnology, Hoban is a member of the Advisory Committee on Agricultural Biotechnology of the U.S. Department of Agriculture (USDA) and recently served on the FDA's biotechnology labeling panel. He is also an advisor to the Council for Biotechnology Information (CBI). Hoban offers a sobering snapshot of the shifting perspectives among leaders in government and industry, who act as public opinion gatekeepers. He believes that the current ambivalence about GM food products expressed by the food industry, wary of a consumer backlash, presents the biggest hurdle to going forward.

C. S. Prakash and Gregory Conko draw on their vast network of international contacts to explain the roots of this sharp divergence in public opinion in "Agricultural Biotechnology Caught in a War of Giants." Prakash is the founder and Conko a board member and vice president of AgBioWorld, the most respected agricultural biotechnology Web site in the world, with endorsements from more than 3,300 scientists in fifty-five countries. The AgBioWorld "Declaration in Support of Agricultural Biotechnology" has been signed by twenty-five Nobel laureates, including Norman Borlaug, James Watson, Arthur Korenberg, Marshall Nirenberg, Peter Doherty, Paul Berg, Oscar Arias, and John Boyer.

Offering a European perspective, Tony Gilland's "Trade War or Culture War? The GM Debate in Britain and the European Union" offers a scathing critique of the lack of resolve of British and EU politicians when faced with an organized antiscience backlash. Based in London, Gilland has long focused on the intersection of science and politics at the British Institute of Ideas, where he has organized provocative conferences on "Genes and Society" and "Interrogating the Precautionary Principle." He has also edited and contributed to numerous books, including *Science—Can We Trust the Experts?* (2002) and *Nature's Revenge?* (2002). Gilland provides evidence that the opposition to biotechnology in Europe is wide but not deep, and flows less from an intrinsic distrust of GM products than from the "growing distrust of political authority and scientific expertise," which he believes are correctable problems.

1

Global Views on Agricultural Biotechnology

Thomas Jefferson Hoban

Conflicts over the use and future of agricultural biotechnology are intensifying between countries that support the development of genetically modified (GM) crops (including the United States, Canada, and many developing countries) and those that oppose such development—mainly the European Union (EU) and Japan. These conflicts are affecting the entire food system from farm to table and presenting challenges for many key stakeholder groups.

Debates have been underway for almost twenty years over the safety of GM crops for the environment and for human health. In addition, biotechnology has become a "lightning rod" for a diverse set of political, social, and economic issues, many of which cannot be resolved by scientific analysis or regulation. These include concerns about corporate control over the food system and the decrease in the number of ag input suppliers (i.e., fewer choices for farmers). Questions are also being raised about the distribution of the benefits and risks of biotechnology.

Clashing cultures and shifting standards have led to a wide range of poorly understood concerns. These conflicts occur because of the way biotechnology affects groups differently. People with adequate food can afford to be increasingly particular about how their food is processed and produced, a trend reflected in the growing interest in organic foods by elite consumers. The poor, however, do not have that choice.

Interest groups have fueled these social conflicts by raising issues that would otherwise not be considered. The organic industry uses consumer anxiety over GM food to win an increasing share of consumers' hearts, minds, and stomachs. Activist groups have launched campaigns against multinational biotechnology companies and industrial agriculture. On the other side, the biotechnology industry and the U.S. government are pushing their own economic and political interests.

At the heart of the public relations battle is the extent to which each opponent develops messages and rationale for its positions. The United States and the EU both want to win support on the African continent. In an attempt to position the EU's cautious stance toward GM food as immoral, the Bush administration, the scientific community, industry groups, and others are making the case that resistance to GM crops inhibits the African countries from accepting U.S. humanitarian aid. Each side claims to have law and/or science on its side.

This conflict reflects significant underlying differences in each country's approach to the evaluation and regulation of technology. In the United States and Canada, private business and government have assumed the stance that a particular technology is safe unless and until it can be proved dangerous. Conversely, European leaders have adopted a version of the precautionary principle that assumes a particular technology is too risky unless and until it can be proved safe.

Many in the EU and elsewhere are not convinced that the safety concerns have been adequately addressed. As the EU labeling policies and procedures become established, it will be necessary to set up complex systems of "identity preservation"—that is, of keeping genetically modified products separated from others—that will add to the price of food, at least in Europe. This will affect consumers in other parts of the world as well, along with key companies from across the food values chain.

Biotechnology has also become a symbol of globalization and a target for growing anti-American sentiment. As trade issues are complicated by the difficulties of segregating grains economically to meet the EU's identity preservation standards, governmental policies and private decisions are implemented in arenas where competing interests must be accommodated. Recently there have been several global government initiatives related to biotechnology that affect the availability of and markets for genetically modified crops. It is important to understand how global policies are created and applied in both developing and industrialized countries.

Food industry opinion leaders also serve as gatekeepers for biotechnology to enter the food system. The food industry plays a vital role in shaping consumers' attitudes and appetites for new food items, including those developed with biotechnology. The world has become one global market for both raw commodities and finished food products, and decisions by food industry gatekeepers have an enormous impact on important stakeholders, from the scientists in the biotech labs to consumers. Market forces have in some cases slowed or stopped development of new crops.

Leaders' Views on Biotechnology: Results of a Survey

A survey research project, completed in the spring of 2002, was designed to ascertain the views of key worldwide opinion leaders representing the food industry and national governments. It assessed stakeholders' attitudes toward the benefits, risks, and regulation of biotechnology, and its findings suggest some strategies for improving communication among all the main stakeholders in this arena.

For the first component, the survey focused on government representatives who had been appointed to represent their countries in one of three international policy arenas. The government officials were drawn from three areas: environmental protection, public health, and agricultural trade. No one from any nongovernment organization or industry was included. Seventy-six of those surveyed were from the thirty member countries of the Organisation for Economic Co-operation and Development (OECD), and 109 were from less-developed, non-OECD countries. These respondents completed and returned a mail questionnaire.

As for the second component, since no publicly available lists are available for the food industry, we had to rely on the advice and assistance from food industry trade associations. Food industry respondents, 241 in all, represented suppliers, manufacturers, distributors, retailers, and restaurants. These were generally from the United States and were interviewed by telephone. For the most part, this chapter will simply compare and contrast the views of the food industry and government leaders. Where this is not the case, it will be clear from the chart and text which of the groups is being discussed.

FIGURE 1-1
LEADERS' VIEWS ON BIOTECHNOLOGY RESEARCH

Question: How important are each of the following research areas related to the use of biotechnology in agriculture and food production? Use a scale from 1 to 5 (where 1 is not at all important, 3 is somewhat important, and 5 is very important).

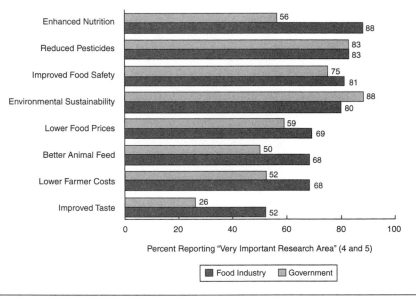

SOURCE: Hoban (2002).

Perceptions of Biotechnology. The food industry and government leaders were asked to rate the importance of a number of different research areas that could be pursued through biotechnology. As figure 1-1 illustrates, both groups saw reduced use of pesticides and environmental sustainability as important research areas. Particularly important for global government leaders were environmental benefits. Food industry leaders were much more enthusiastic than global representatives in their support for research to enhance nutrition, lower costs, improve animal feed, and improve taste, benefits that will be important for developing countries.

As illustrated in figure 1-2, the government leaders expressed the most concern over the uncertainty associated with any long-term effects

FIGURE 1-2

GOVERNMENT LEADERS' CONCERNS ABOUT BIOTECHNOLOGY

Question: *How serious are each of the following concerns that have been raised about the use of biotechnology in agriculture and food production? Use a scale from 1 to 5 (where 1 is not at all serious, 3 is somewhat serious, and 5 is very serious).*

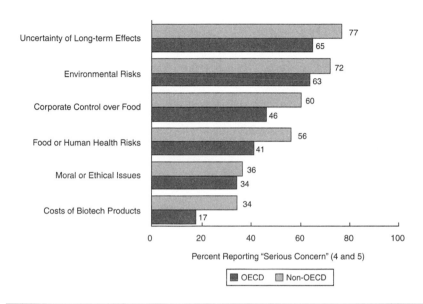

SOURCE: Hoban (2002).

of biotechnology. Almost as many were concerned about environmental or ecological risks (significantly more than were concerned about food safety and health). Only a third expressed concern over moral and ethical issues. The least concern was expressed about the costs of biotech products. In all cases, government representatives from non-OECD (i.e., less developed) countries expressed greater concern about the potential for problems.

Results from the food industry leaders, presented in figure 1-3, are based on open-ended questions in which respondents were asked to name the most serious concerns their company had about biotechnology. This was followed by a question asking what respondents thought consumers were most concerned about. Over three-quarters of respondents

FIGURE 1-3
FOOD INDUSTRY LEADERS' CONCERNS ABOUT BIOTECHNOLOGY

Question: *What is the most important concern your company has [consumers have] about the use of biotechnology in agriculture and food production?*

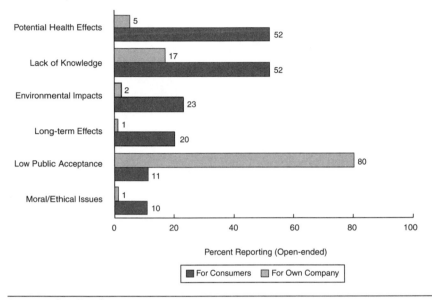

SOURCE: Hoban (2002).

indicated that their companies were most concerned with lack of consumer acceptance. Another 17 percent indicated that lack of knowledge was their most serious problem. When asked about consumers, most food industry leaders mentioned concerns over potential health effects and an overall lack of knowledge about food biotechnology.

Another important area involves the distribution of benefits from agricultural biotechnology. As shown in figure 1-4, over 80 percent of both groups indicated that the biotechnology companies would be receiving most of the benefits. It is interesting that government leaders are much more likely than the food industry respondents to anticipate consumer benefits from biotechnology. One clear issue of concern for the future is that the food industry generally does not foresee any benefits from the use of biotechnology.

FIGURE 1-4

LEADERS' ASSESSMENT OF HOW MUCH DIFFERENT STAKEHOLDER GROUPS
WILL BENEFIT FROM THE USE OF BIOTECHNOLOGY

Question: *How much do you think each of the following groups will benefit from the use of biotechnology in agriculture and food production? Please use a scale from 1 to 5 where 1 means they will not benefit at all, 3 means they will receive some benefit, and 5 means they will receive a lot of benefit.*

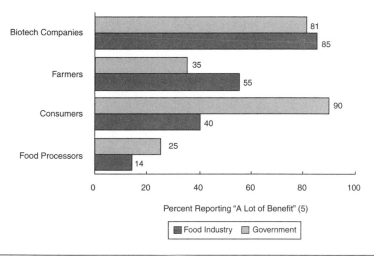

Percent Reporting "A Lot of Benefit" (5)

■ Food Industry ▨ Government

SOURCE: Hoban (2002).

Communication Challenges and Opportunities. Improved communication will play an important role in making sure that the benefits of biotechnology are achieved, while keeping risks and negative socioeconomic impacts to a minimum. What types of information will stakeholders need? As shown in figure 1-5, food industry leaders were most interested in information about the impacts of biotechnology on food safety and nutrition, consumer acceptance issues, effects on the environment, and effects on food quality and taste. Global government leaders were particularly concerned about the impacts of biotechnology on food safety and nutrition and the environment. Neither group was very interested in basic scientific information or effects on costs.

According to figure 1-6, leaders from both sectors reported the most trust in university scientists and medical associations. Not as many would

FIGURE 1-5

LEADERS' RATINGS OF IMPORTANCE FOR ADDITIONAL BIOTECHNOLOGY
INFORMATION

Question: *How important would it be for you to receive more information about each of the following? Use a scale of 1 to 5 (where 1 is not at all important, 3 is somewhat important, and 5 is very important).*

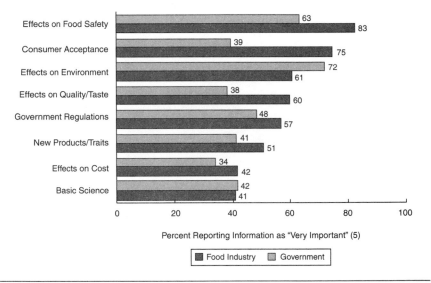

Percent Reporting Information as "Very Important" (5)

■ Food Industry   ☐ Government

SOURCE: Hoban (2002).

have high trust in farmers or the biotechnology companies. Both food industry and government representatives reported relatively low trust in consumer groups (especially Greenpeace). Put another way, respondents reported the most trust in information from third-party sources (such as university scientists and health professionals), but had much less confidence in the interest groups on either side of the debate.

Societal Acceptance of Biotechnology. Respondents were asked to rate the effectiveness of various ways of building consumer confidence in biotechnology. As shown in figure 1-7, the food industry was most supportive of increased consumer education. They also recognized the importance of science-based regulations and products with clear consumer

FIGURE 1-6

FOOD INDUSTRY AND GOVERNMENT LEADERS' TRUST IN KEY
BIOTECHNOLOGY STAKEHOLDERS

Question: *Suppose a number of groups made public statements about the safety of foods developed through biotechnology. Would you have a lot, some, or no trust in statements made by . . . ? Use a scale of 1 to 5 (where 1 is no trust, 3 is some trust, and 5 is a lot of trust).*

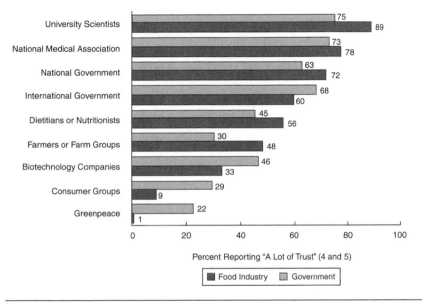

Percent Reporting "A Lot of Trust" (4 and 5)

■ Food Industry   ▨ Government

SOURCE: Hoban (2002).

benefits. Global government leaders indicated that conducting more research on the environmental implications of biotechnology would be the most effective means of building consumer confidence, and they recognized the importance of consumer education and more transparent decision-making processes.

Looking again at just the food industry leaders, respondents were asked to evaluate the importance of eight possible reasons the EU had been more negative about biotechnology than the United States. In figure 1-8, two reasons stand out as most important: negative media coverage and previous food safety scares. Other important issues included lack of

FIGURE 1-7

EFFECTIVENESS OF ALTERNATIVE WAYS TO BUILD PUBLIC
CONFIDENCE IN BIOTECHNOLOGY

Question: *Suppose that someone was interested in how to build consumer confidence in food and agricultural biotechnology. How much would each of the following build consumer confidence in modern biotechnology?*

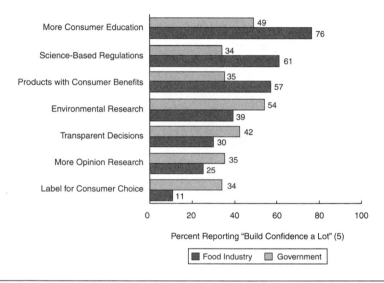

Percent Reporting "Build Confidence a Lot" (5)

■ Food Industry   ▨ Government

SOURCE: Hoban (2002).

confidence in government agencies and the fact that biotech opponents had been much more successful in Europe. On the other hand, relatively few food industry leaders felt that different cultural views on food or supermarkets' reactions to biotechnology were important reasons for the differences between the United States and the EU.

Finally, government policies are important for ensuring that the products of biotechnology are safe for human health and the environment. Respondents in each group were asked to rate three different government entities in terms of confidence in each government's capabilities. As shown in figure 1-9, by far, the food industry leaders expressed the greatest confidence in the U.S. government's approach and capabilities relative to biotechnology regulation. They had the least confidence in the European

FIGURE 1-8

FOOD INDUSTRY LEADERS' VIEWS ON WHY EUROPE HAS BEEN MORE NEGATIVE THAN THE UNITED STATES TOWARD BIOTECHNOLOGY

Question: *As you may know, consumers in some countries (such as Great Britain) have been more negative about modern biotechnology than consumers in other countries (such as the United States). Tell me how important each of the following reasons is for explaining why consumers in some countries are more negative.*

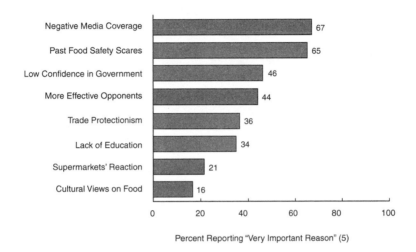

Percent Reporting "Very Important Reason" (5)

SOURCE: Hoban (2002).

approach to regulation. Responses of the global government representatives were the opposite, with greatest trust shown in the EU's more precautionary approach.

Conclusions

Food industry leaders have been supportive, while government representatives have been less comfortable with food biotechnology. Government representatives from non-OECD countries have been much more likely to express concerns about the risks of biotechnology than OECD government

FIGURE 1-9

**LEADERS' CONFIDENCE LEVELS IN DIFFERENT GOVERNMENT BODIES TO
REGULATE BIOTECHNOLOGY EFFECTIVELY**

Question: *Briefly consider the role of government in biotechnology. Do you have a lot,
some, or no confidence in . . . to effectively regulate biotechnology?*

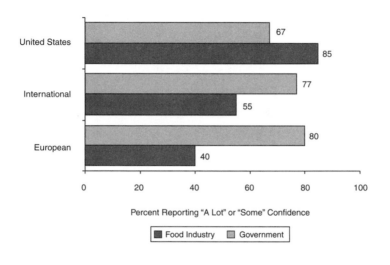

Source: Hoban (2002).

representatives. Food industry and government representatives have agreed
that increasing crop yields and enhancing nutrition are important priorities
for farmers and the food industry.

Both food industry and global government representatives appear to
agree that consumer education is critical to market acceptance of foods
produced through biotechnology. The key is to build understanding of
biotechnology's safety and benefits, while also acknowledging the risks.
Opinion leaders will ultimately be responsible for communicating with
the public in particular countries, so it is vital they receive balanced information in a timely and credible manner. We need to enhance communication and ensure mutual benefits for all parties in the food value chain,
from technology providers to farmers, food handlers and processors,
retailers, and food service providers.

Developing countries (non-OECD) are in a key position to influence the future acceptance of agricultural biotechnology. They want and deserve to make their own decisions without pressure from either the United States or the EU. The developing world needs assistance and infrastructure from the developed world to utilize tools of modern biotechnology. Here again it will be important for government decision makers to have the most credible and accurate information available.

As with any innovation, the use of modern biotechnology has the potential for both benefits and risks. The challenge is to ensure that the risks are kept to a minimum while the benefits become widely available. There is no such thing as a risk-free technology. The benefits of innovation must be weighed against the consequences of maintaining the status quo. It is also important to compare the risks of a new technology to the approaches that are already in use. The more we use biotechnology and the more complicated it gets, as in the fields of biopharming and transgenic animals, the greater the risks will be.

As the technology jumps from the fairly simple process of adding a single gene to a plant to the more complex one of essentially reshaping plants at will, everyone involved will need to be more open with the public. The first crops were designed to benefit farmers by saving them money and time. Emerging products will have direct impacts on people. We need more regulation—not less—of those that are designed to be active in the human body.

Some outstanding issues need to be addressed. Most Americans still believe they are not eating GM food. As it shifts the way GM ingredients and foods are evaluated, the food industry needs to "break the news" about the widespread presence of GM ingredients. Currently, GM ingredients are evaluated in terms of "substantial equivalence." According to this concept, if a new food or food component is found to be substantially equivalent to an existing one, it can be treated in the same manner with respect to safety—that is, it can be concluded to be as safe as the conventional one. However, the industry is moving to foods that will be functionally active in the human body (and therefore no longer "substantially equivalent"). In the future there will need to be much more stringent testing and regulation of foods that contain enhanced nutrition or other substantial changes.

The research presented here suggests ways to improve public understanding of biotechnology. The first need is for greater communication among all groups. Universities and other third parties must disseminate balanced information to all stakeholders through a variety of communication channels. It is also important to improve and maintain the credibility of the government regulatory systems. This requires greater transparency and a real commitment to involving all interested and affected parties.

Ultimately, it will be important to deliver products that have benefits for all stakeholder groups. More research needs to be focused on fruits and vegetables that provide real benefits to consumers, such as enhanced nutrition, improved taste, and extended shelf life. New products with direct benefits to food processors and others along the food value chain need to be introduced into the market soon. Steps also need to be taken to develop more cost-effective and efficient identity preservation systems so consumers can avoid GM foods if they prefer. Such systems are needed to ensure that specialty products can be kept separate and command a premium price for the producers.

Substantial trade disagreements involving biotechnology in food and agriculture still exist between the United States and the EU. In the aftermath of the Iraqi invasion, anti-American sentiment is at a historic high. There are also strong economic interests on both sides of this debate. An acceptable resolution is unlikely unless the parties view the dispute as a sociocultural conflict rather than a disagreement over scientific facts.

The results of the leaders' survey suggest how to move forward toward a global consensus. The biotechnology industry, farmers, and other biotech proponents need to accept the fact that "sound science" is only one criterion for public policymaking. For a growing number of people, particularly in Europe, science alone is not persuasive. People make decisions based more on emotion than logic, especially with regard to food. They also question the moral and ethical aspects of biotechnology. The only viable approach is to win consumers' hearts, minds, and stomachs, rather than try to force-feed them. The United States could win the trade battle but lose the consumer acceptance war.

Perception is even more important than reality when it comes to social acceptance of GM foods. Industry's simplistic reliance on education about biotech benefits will not be enough to calm deep-seated anxieties. So far, neither consumers nor the food industry have seen any real benefits from biotechnology. Many people have concerns that go beyond risks to human health or the environment. They believe that benefits are unlikely to out-weigh moral or ethical objections.

The fair distribution of costs and benefits is a key ethical issue that needs to be addressed. All parties need to discuss and debate who wins and who loses before products are commercialized. Like other innovations, biotechnology will have short- and long-term consequences. Anticipating potential impacts will inform public debate and policymaking, as well as minimize potentially disruptive effects.

Reference

Hoban, Thomas. 2002. *Global Opinion Leader Survey: Perspectives on Biotechnology*. Raleigh: North Carolina State University.

2

Agricultural Biotechnology Caught in a War of Giants

C. S. Prakash and Gregory Conko

To deny desperate, hungry people the means to control their futures by presuming to know what is best for them is not only paternalistic but morally wrong. . . . We want to have the opportunity to save the lives of millions of people and change the course of history in many nations. . . . The harsh reality is that, without the help of agricultural biotechnology, many will not live.

—Hassan Adamu, Nigerian minister of agricultural and rural development (2000)

Farmers in the United States have reaped many benefits from bioengineered crops, but the technology holds even greater promise for farmers in less-developed countries. Many biotechnology products currently in the research pipeline are being developed specifically for resource-poor farmers, and some of the varieties that have proved a boon to industrialized nations are already helping farmers in less-developed nations. To make those benefits more widely available, political obstacles must first be overcome, but the technology has been demonstrated to be safe for consumers and the environment, justifying widespread use.

Today, most people around the world have access to a greater variety of nutritious and affordable foods than ever before, thanks mainly to developments in agricultural science and technology. The average human lifespan—arguably the most important indicator of quality of life—has increased steadily in the past century in almost every country. Even in many

35

less-developed countries, the average lifespan has doubled in the past few decades. Despite the massive increase in population, the global malnutrition rate has decreased in that period from 38 percent to 18 percent, and giants such as India and China have quadrupled their food-grain production.

The record of agricultural progress during the past century speaks for itself. Those countries that developed and embraced improved farming technologies have brought unprecedented prosperity to their people, made food much more affordable and abundant, helped to stabilize farm yields, and reduced the destruction of wild lands (Borlaug 2000). The productivity gains derived from scientifically bred, high-yielding crop varieties, as well as improved use of synthetic fertilizers and pesticides, allowed the world's farmers to double global food output during the past fifty years on roughly the same amount of land, at a time when global population rose more than 80 percent (Goklany 1999). Without these improvements in plant and animal genetics and other scientific developments, collectively known as the Green Revolution, we would today be farming on every square inch of arable land to produce the same amount of food, destroying hundreds of millions of acres of pristine wilderness in the process.

The Challenge of Food Security in Developing Countries

Many less-developed countries in Latin America and Asia have benefited tremendously from the Green Revolution. But due to a variety of factors, both natural and human, agricultural technologies have not spread equally across the globe. Many people in sub-Saharan Africa and parts of South Asia continue to suffer from abject rural poverty driven by poor farm productivity. Some 740 million people go to bed daily on empty stomachs, and nearly 40,000 people—half of them children—die every day from hunger- or malnutrition-related causes. Unless trends change soon, the number of undernourished could well surpass 1 billion by 2020 (International Food Policy Research Institute 2001).

What can we do about this needless and cruel situation? In many less-developed countries, where subsistence farmers eke out meager livings and the ability to provide enough food for mere survival is perilously uncertain, the importance of increasing the yields of staple crops such as rice, millet, cowpea, sweet potato, and cassava cannot be overstated. In many places, the loss of a crucial crop to pests, diseases, or weather can literally mean the difference between life and death, threatening the well-being of entire communities.

Despite assertions by critics of technology, providing genuine food security for such people must include solutions other than mere redistribution. According to a report published jointly by the National Academies of Science from Brazil, China, India, Mexico, and the United States, the United Kingdom's Royal Society, and the Third World Academy of Science,

> In developing countries . . . about 650 million of the poorest people live in rural areas where the local production of food is the main economic activity. Without successful agriculture, these people will have neither employment nor the resources they need for a better life. . . . Farming the land is the engine of progress . . . in less developed countries. (Royal Society of London et al. 2000, 4)

Coupled with that great need is the fact that the rate of increase in food production globally dropped from 3 percent per annum in the 1970s to 1 percent per annum by the late 1990s (Conway and Toennissen 1999, C56). The needs of burgeoning populations, especially in the developing world, could very well soon outstrip global food production. The UN Food and Agriculture Organization (FAO) expects the world's population to grow to more than eight billion by 2030 and projects that global food production must increase by 60 percent to accommodate the estimated population growth, close nutrition gaps, and allow for dietary changes over the next three decades (Bruinsma 2003).

Modern Biotechnology Joins Crop Development

Although better farm machinery and development of synthesized fertilizers, insecticides, and herbicides have been extremely useful, an improved understanding of the principles of inheritance and modern genetics has been the single most important factor in enhancing food production (*AgNews* 2003). Every crop is a product of repeated genetic editing by humans over the past few millennia. Our ancestors chose a few once-wild plants and gradually modified them simply by selecting those with the largest, tastiest, or most robust offspring for propagation. In that way, organisms have been altered so greatly that traits present in existing populations of cultivated rice, wheat, corn, soy, potatoes, tomatoes, and many others have very little in common with their ancestors. Wild tomatoes and potatoes contain very potent toxins, for example, but today's cultivated varieties have been modified to produce healthy and nutritious food.

Hybridization—the mating of different plants of the same species—has helped us assimilate desirable traits from several varieties into elite cultivars. And when desired characteristics have been unavailable in cultivated plants, genes have been liberally borrowed from wild relatives and introduced into crop plants. Commercial tomato plants are commonly bred with wild tomatoes to introduce improved resistance to pathogens, nematodes, and fungi (Bruening 2002). Successive generations then have to be carefully "back-crossed" into the commercial cultivars to eliminate any unwanted traits accidentally transferred from the wild varieties, such as the glyco-alkaloid toxins common in tomatoes and potatoes.

Even when crop and wild varieties refuse to mate, various tricks can be employed to produce "wide crosses" between two plants that are otherwise sexually incompatible. Often, the embryos created by wide crosses die prior to maturation, so they must be "rescued" and cultured in a laboratory environment. Even then, the rescued embryos typically produce sterile offspring. They can only be made fertile again by using mutagenic chemicals that cause the plants to produce a duplicate set of chromosomes (Goodman et al. 1987, 52). The plant triticale, an artificial hybrid of wheat and rye, is one such example of a wide-cross hybrid made possible solely by the existence of embryo-rescue and chromosome-doubling techniques. Triticale is now grown on over three million acres worldwide, and dozens of other products of wide-cross hybridization are also common.

When a desired trait cannot be found within the existing gene pool, breeders can create new variants by intentionally mutating plants with x-ray or gamma radiation or with mutagenic chemicals, or simply by culturing clumps of cells in a petri dish and allowing mutations to occur spontaneously as the cell culture grows. For example, ClearField wheat, a relatively new mutant variety commercialized in the United States and Australia, was produced with chemical mutation to be resistant to the BASF herbicide Beyond. Mutation breeding has been in common use since the 1950s, and more than 2,250 known mutant varieties have been bred in at least fifty countries, including France, Germany, Italy, the United Kingdom, and the United States (Maluszynski et al. 2000).

While government regulators and environmental activists consider wide crosses and mutation techniques to be conventional forms of breeding, they involve gross genetic manipulation and can occur only in a laboratory environment. Recombinant DNA methods—what are generally considered to be "biotechnology"—thus must be seen as a recent extension of the continuum of techniques we have employed to modify and improve our crops. The primary difference is that modern bioengineered crops involve a precise transfer of one or two known genes into plant DNA—a surgical alteration of a tiny part of the crop's genome. Compare this to the sledgehammer approaches of traditional and wide-cross hybridization or mutation breeding, which bring about gross genetic changes, many of them unknown and unpredictable. Furthermore, unlike varieties developed from more conventional breeding, modern bioengineered crops are rigorously tested and subjected to intense regulatory scrutiny before being commercialized (Prakash 2001, 13).

The scientific community, plant breeders, and farmers have thus overwhelmingly accepted and supported the use of gene technologies in crop improvement. Accumulated knowledge after decades of plant breeding, combined with expert judgment, science-based reasoning, and empirical research, has further generated confidence that modern bioengineered crops pose no new or heightened risks which cannot be identified and mitigated, and that any unforeseen hazards are likely to be negligible and manageable.

Leading scientists around the world have attested to the health and environmental safety of agricultural biotechnology, and they have called for bioengineered crops to be extended to those who need it most: hungry people in the developing world. The technology has been endorsed by dozens of scientific and health associations, including the U.S. National Academy of Sciences (1987), the United Kingdom's Royal Society (Royal Society of London et al. 2000), the United Nations Development Programme (2001), and many others. Nearly 3,500 eminent scientists from all around the world, including twenty-five Nobel laureates, have signed a declaration supporting the use of agricultural biotechnology (AgBioWorld Foundation 2005). The scientific literature is filled with hundreds of peer-reviewed studies describing the safety of bioengineered crops and foods. And a review of eighty-one separate research projects, conducted over fifteen years and funded by the European Union, found that bioengineered crops and foods are just as safe for the environment and for human consumption as conventional crops, and in some cases even safer, because the genetic changes in the plants are much more precise. This confidence has been validated by the excellent safety record of the biotech crops and the food derived from such crops since their commercial inception almost a decade ago (Kessler and Economidis 2001).

When given a choice, most farmers choose biotech seeds because of the value they bring. Farmers around the world have embraced modern genetic technology because it makes farming more efficient, protects or increases yields, and reduces reliance on chemicals that, other things being equal, they would prefer not to use. Opponents of the technology like to discount or even ignore such information as corporate public relations puffery. However, crops enhanced through modern biotechnology are now grown on nearly 200 million acres (81 million hectares) in seventeen countries. More importantly, 90 percent of the 8.25 million growers who have benefited from bioengineered crops were resource-poor farmers in the developing world (James 2004).

Unremarkably, most commercially available biotech plants were designed for farmers in the industrialized world. They include varieties of corn, soybean, potato, and cotton modified to resist insect pests and plant diseases, and to make weed control easier. However, the increasing adoption of transgenic varieties by developing countries over the past few years has shown that they can benefit at least as much as, if not more than, their counterparts in industrialized countries. The productivity of farmers everywhere is limited by crop pests and diseases—and these are often far worse in tropical and subtropical regions than in the temperate zones.

About 20 percent of plant productivity in the industrialized world—but up to 40 percent in Africa and Asia—is lost to insect pests and pathogens, despite the ongoing use of copious amounts of pesticides (Sommerville and Briscoe 2001, 2217). The European corn borer destroys approximately

7 percent, or 40 million tons, of the world's corn crop each year—a sum equivalent to the annual food supply, in calories, for 60 million people. It should come as no surprise that when permitted to grow bioengineered varieties, poor farmers in less-developed nations have eagerly snapped them up. According to the International Service for the Acquisition of Agri-Biotech Applications, farmers in less-developed countries now grow nearly one-third of the world's transgenic crops on more than 62 million acres (27.6 million hectares), and they do so for many of the same reasons that farmers in industrialized nations do (James 2004).

But the benefits of these traits do not end there. It should be relatively easy to move insect- and disease-resistant traits into crops such as cassava and sweet potato, staples of the central African diet that provide vital sources of calories and essential vitamins and minerals to millions. In 1998, the people of Africa lost 60 percent of their cassava crop to mosaic virus. Sweet potato yields in many African nations have been lain dangerously low, with some farmers losing up to 80 percent of expected yields to the sweet potato weevil and plant viruses. Biotech scientists are working to solve these problems by producing plants that resist pests and disease, major causes of crop damage in the developing world. Research is under-way on sweet potatoes that produce their own protection against diseases and insects, as well as beans, cassava, and other staple foods with enhanced natural tolerance to diseases, pests, and physical stresses.

In 1997, the World Bank's Consultative Group on International Agricultural Research estimated that biotechnology could help increase world food production by up to 25 percent (Kendall et al. 1997). "The need for biotechnology in Africa is very clear," noted the late John Wafula, head of biotechnology research at the Kenya Agricultural Research Institute. "The use of high-yielding, disease-resistant and pest-resistant crops would have a direct bearing on improved food security, poverty alleviation, and environmental conservation in Africa" (Mgendi 1999).

Biotechnology Has Begun to Affect the Developing World

World hunger is a complex problem for which there is no single solution. Biotechnology will not, by itself, conquer it, but it can be a valuable tool in the struggle. South African and Chinese farmers have been growing transgenic, insect-resistant cotton for several years, achieving yield increases of more than 20 percent and cost savings of nearly 30 percent (Ismael, Bennett, and Morse 2001, 18; Huang et al. 2002, 676). The Indian government approved insect-resistant cotton for commercial cultivation in the spring of 2002. This cotton variety, similar to those so popular in the United States, boosted yields by 30 percent or more in field trials conducted by Indian farmers (Qaim and Zilberman 2003, 901). Eventually, it could transform India from the world's third-largest producer of cotton into the largest. Since cotton is not a food crop, the governments of many less-developed countries have felt more comfortable approving it as their first bioengineered plant because it sidesteps global restrictions on bioengineered food imports. But cotton is an important cash crop, and bioengineered varieties generate important financial gains for growers.

In addition, South African and Philippine farmers are now growing transgenic corn, and South Africans are growing transgenic soybeans, which have proved quite popular in South America as well. Farmers in Argentina, the world's third-largest producer of soybeans, plant more than 90 percent of their soybean acreage with bioengineered varieties (White 2001). Experience with the product there has been so positive that neighboring farmers in Paraguay, where no biotech varieties have yet been approved by the government, and Brazil, where the crop was not approved until 2003, have smuggled large amounts of bioengineered soybean seed from Argentina to grow the variety illegally (Blackburn 2003; Commission of the European Communities 2000).

Aside from improving productivity, the use of bioengineered plants could result in a dramatic reduction in dangerous pesticide spraying. In China, for example, where pesticides are typically sprayed on crops by hand, some four to five hundred cotton farmers die every year from acute pesticide poisoning. A study conducted by researchers at Rutgers University in the United States and the Chinese Academy of Sciences found that the adoption of transgenic cotton varieties in China has lowered the amount of pesticides used by more than 75 percent and reduced the number of pesticide poisonings by an equivalent amount (Huang et al. 2002). Another study, by economists at the University of Reading in Britain, found that South African cotton farmers have seen similar benefits (Ismael, Bennett, and Morse 2001, 16; Kirsten and Gouse 2002).

The reduction in pesticide spraying also means that fewer natural resources are consumed to manufacture and transport the chemicals. Researchers in the United States found that in 2000 alone, farmers growing transgenic cotton used 2.4 million fewer gallons of fuel and 93 million fewer gallons of water, and were spared some 41,000 ten-hour days needed to apply pesticide sprays (Falck-Zepeda et al. 2000). The National Center for Food and Agricultural Policy further found that biotechnology-derived plants—soybeans, corn, cotton, papaya, squash, and canola—increased U.S. food production by approximately four billion pounds, saved $1.2 billion in production costs, and decreased pesticide use by about 46 million pounds in 2001 alone (Giannessi 2002, 8–9).

Soon many transgenic varieties that have been created specifically for use in less-developed nations will be ready for commercialization. Examples include insect-resistant rice varieties for Asia, pest-resistant vegetable crops in India and China, and virus-resistant papaya for Caribbean nations. The next generation of transgenic crops now in research labs around the world is poised to bring even further productivity improvements for the poor soils and harsh climates that are characteristic of impoverished regions (Conway and Toennissen 1999, C56). Scientists have already identified genes for resistance to environmental stresses common in tropical nations, including tolerance to soils that have high salinity or are particularly acidic or alkaline.

A Mexican research group, led by the highly regarded Luis Herrera-Estrella, has shown that tropical crops can be modified using biotechnology to tolerate acidic soils better, significantly increasing the productivity of corn, rice, and papaya (Manuel de la Fuente et al. 1997). These traits for greater tolerance to adverse environmental conditions would be tremendously advantageous to poor farmers in less-developed countries, especially those in Africa. By making it easier to farm marginal lands and improving productivity, the planting of hardier crops may also help to conserve hotspots of biodiversity such as wetlands and tropical rainforests by reducing or eliminating the need to convert undeveloped land to food production (Conko 2003; Conko and Prakash 2002).

The primary reason Africa never benefited from the Green Revolution is that plant breeders focused on improving crops such as rice, wheat, and corn, which are not widely grown there. Plus, much of the African dry lands have little rainfall and no potential for irrigation, both of which play essential roles in productivity success stories for crops such as Asian rice. The remoteness of many villages and the poor transportation infrastructure in landlocked African countries would make it difficult for farmers to obtain agricultural chemicals, such as fertilizers, insecticides, and herbicides, even if they could be donated by aid agencies and charities, or if the farmers had the money to purchase them.

Many of these obstacles could be overcome by biotechnology. Packaging technologies within seeds could provide the same, or better, productivity advantage as chemical or mechanical inputs, but in a much more user-friendly manner. Farmers could control insect pests, viral or bacterial pathogens, extremes of heat or drought, and poor soil quality just by planting these crops.

Still, antibiotechnology activists like Vandana Shiva, of the Research Foundation for Science, Technology, and Ecology, based in New Delhi, and Miguel Altieri, of the University of California at Berkeley, argue that poor farmers in less-developed nations will never benefit from biotechnology because it is controlled by multinational corporations. According to Altieri, "Most innovations in agricultural biotechnology have been profit-driven rather than need-driven. The real thrust of the genetic engineering industry is not to make third world agriculture more productive, but rather to generate profits" (Altieri and Rossett 1999, 157). But that sentiment is not shared by the thousands of academic and public-sector researchers actually working on biotech applications in those countries. Cyrus Ndiritu, former director of the Kenyan Agricultural Research Institute, argues, "It is not the multinationals that have a stranglehold on Africa. It is hunger, poverty and deprivation. And if Africa is going to get out of that, it has to embrace [biotechnology]" (Wambugu 2001, 5).

Improved Health through Better Nutrition

In addition to increasing productivity, biotechnology offers hope of improving the nutritional benefits of many foods. The next generation of bioengineered products now in development at labs around the world is poised to bring direct health benefits to consumers through enhanced nutritive qualities that include higher protein content, better-quality proteins, lower levels of saturated fat, increased vitamins and minerals, and many others (Pew Initiative on Food and Biotechnology 2001). Bioengineering can also reduce the level of natural toxins (such as the cyanogens in cassava and phytohemagglutinin in kidney beans) and eliminate certain allergens from foods like peanuts, wheat, and milk (American Medical Association 2001). Many of these products are being developed primarily or even exclusively for poor subsistence farmers and consumers in less-developed countries.

Among the best-known bioengineered food products is Golden Rice, which is genetically enhanced with added beta carotene that is converted to vitamin A in the human body. Another variety of rice developed by the same research team has elevated levels of digestible iron (Ye et al. 2000, 304). The diet of more than three billion people worldwide includes inadequate levels of such essential vitamins and minerals. Deficiency in just these two micronutrients can result in severe anemia, impaired intellectual development, blindness, and even death (Della Penna 1999, 377). And even though charities and aid agencies such as the United Nations Children's Fund and the World Health Organization have made important strides in reducing vitamin A and iron deficiency, success has been fleeting. No permanent effective strategy has yet been devised, but Golden Rice may finally provide one.

Importantly, the Golden Rice project is a prime example of the value of extensive public-sector and charitable research activities. The development of the rice was funded mainly by the Rockefeller Foundation in New York, which promised to make it available to poor farmers at little or no cost. It was created by scientists at public universities in Switzerland and Germany, with assistance from the International Rice Research Institute, based in the Philippines, and several multinational corporations (Rockefeller Foundation 1999).

Scientists at publicly funded, charitable, and corporate research centers are developing many similar crops as well. In 2003, Indian scientists announced they would soon make a new high-protein potato variety available for commercial cultivation (Irish Examiner 2003). Another team of Indian scientists, working with technical and financial assistance from Monsanto, is developing an improved mustard variety enhanced with higher levels of beta carotene and other carotenoids in its cooking oil. One lab at

Tuskegee University is enhancing the level of dietary protein in sweet potatoes, a common staple crop in sub-Saharan Africa. Researchers are also developing varieties of cassava, rice, and corn that more efficiently absorb trace metals and mineral nutrients from the soil, have enhanced starch quality, and contain more beneficial vitamins and other micro- and macronutrients than existing varieties.

Research is already underway on fruits and vegetables that could one day deliver lifesaving vaccines. One prospect is a banana that could deliver the vaccine for hepatitis B; another is a potato that provides immunization against diarrheal diseases. These would make inoculation against deadly diseases possible by using locally grown crops that are easy to handle, distribute, and administer (Langridge 2000).

Admittedly, experts recognize that the problem of hunger and malnutrition is not caused solely by a shortage of food. The primary causes of endemic hunger in the countries where it exists have been political unrest and corrupt governments, poor transportation and infrastructure, and, of course, poverty. All of these problems and more must be addressed if we are to ensure real, worldwide food security. But producing enough for 8 or 9 billion people will require greater yields in the regions where food is needed most, and from crops consumed by resource-poor people. New varieties derived through biotechnology provide good, low-input tools to achieve this goal.

High Anxiety—Public Unease over Bioengineered Crops?

Ingredients produced from modern biotech modification are found in thousands of food products consumed worldwide (Bruening 2002). Yet, even though no legitimate evidence of harm to human health or the environment from these foods is known or expected, there is an intense debate questioning the value and safety of bioengineered organisms. Although it may seem reasonable for consumers to express a concern that they "don't know what they're eating with bioengineered foods," it must be reiterated that consumers never had that information with conventionally modified crops, either.

While no assurance of perfect safety can be made, breeders know far more about the genetic makeup, product characteristics, and safety of every modern bioengineered crop than those of any conventional variety ever marketed. Breeders know exactly what new genetic material has been introduced. They can identify where the transferred genes have been inserted into the new plant. They can test to ensure that the transferred genes are working properly and that the nutritional elements of the food have been unchanged. None of these safety assurances has ever before been made with conventional breeding techniques.

Consider, for example, how conventional plant breeders would develop a disease-resistant tomato. Sexual reproduction introduces chromosome fragments from a wild relative in the hope of transferring a gene for disease resistance into cultivated varieties. In the process, hundreds of unknown and unwanted genes are also introduced, with the risk that some of them could encode toxins or allergens, which are known to be present in wild tomatoes (Bruening 2002). Yet regulators never routinely test conventionally bred plant varieties for food safety or environmental risk factors, and their production is subject to practically no government oversight.

We have always lived with food risks. But modern genetic technology makes it increasingly easier to reduce those risks. Societal anxiety over the new genetic modification is, in some ways, understandable. It is fueled by a variety of causes, including unfamiliarity, lack of reliable information about scientific and regulatory safeguards, a steady stream of negative opinion in the news media, opposition by activist groups, growing mistrust of industry, and a general lack of awareness of how our food production system evolved over time. But saying that public apprehension over biotechnology is understandable is not the same as saying that it is valid. There is no evidence at all that bioengineering creates any new environmental or public health risks.

Do Bioengineered Crops Harm the Environment?

Certain aspects of all farming have a negative impact on biodiversity and on air, soil, and water quality. But do modern bioengineered crops pose greater environmental risks, as critics claim? In fact, the innovation of biotechnology has proved safer for the environment than anything dating after the invention of the plow.

The risk of cross-pollination from crops to sexually compatible wild relatives has always existed, and such "gene flow" occurs whenever they grow in close proximity. Moreover, breeders have continuously introduced genes for disease and pest resistance through conventional breeding into all of our crops. Traits such as stress tolerance and herbicide resistance have also been introduced in some crops with conventional techniques, and the growth habits of every crop have been altered. Thus, not only is gene modification a common phenomenon, but so are many of the specific kinds of changes made with recombinant DNA techniques (Prakash 2001).

Naturally, with both conventional and rDNA-enhanced breeding, we must be vigilant to ensure that newly introduced plants do not become invasive and that weeds do not become noxious because of genetic modification. Although modern genetic modification expands the range of new traits that can be added to crop plants, it also ensures that more will be known about those traits, and that the behavior of the modified plants will, in many ways, be easier to predict. That greater knowledge, combined with historical experience with conventional genetic modification, provides considerable assurance that such risks will be minimal and manageable.

No major problems with weeds or invasiveness have developed since the advent of modern plant breeding, because domesticated plants are typically poorly fit for survival in the wild. If this were a genuine problem, it would have emerged centuries ago. Concerns about bioengineered crops running amok, or errant genes flowing into wild species—sometimes characterized as "gene pollution"—pale in comparison to the genuine risk posed by intentionally or unintentionally introducing totally unmodified "exotic" plants into new ecosystems. Notable examples of the latter include water hyacinth in Lake Victoria, cordgrass in China, cattail in Nigeria, and kudzu in North America (Bright 1999, 54–55).

This is not, of course, to say that no harm could ever come from the introduction of modern bioengineered or conventionally modified crop varieties. Some traits, if transferred from crops to wild relatives, could increase the reproductive fitness of weeds and cause them to become invasive or erode the genetic diversity of native flora, and a few cases of this happening have been documented with conventionally bred crops (Rissler and Mellon 1996). But the magnitude of that risk has solely to do with the traits involved, the plants into which they are transferred, and the environment into which they are introduced. Consequently, breeders, farmers, and regulators are aware of the possibility that certain traits introduced into any new crop varieties, or new varieties introduced into different ecosystems, could pose genuine problems, and these practices are carefully scrutinized. Again, this risk occurs regardless of the breeding method used.

Finally, one must also recognize the potential positive impact of rDNA-modified crops on the environment. Already, commercialized bioengineered crops have reduced agricultural expansion and promoted ecosystem preservation. They have improved air, soil, and water quality as a consequence of reduced tillage, chemical spraying, and fuel use, and they have enhanced biodiversity because of lower insecticide use (Carpenter et al. 2002).

We have already noted that the growing of modern bioengineered crops has significantly reduced pesticide use in the United States. A 2002 Council for Agricultural Science and Technology report also found that rDNA-modified crops promoted the adoption of conservation tillage practices, resulting in many other important environmental benefits: 37 million tons of topsoil preserved, 70 percent reduction in herbicide runoff, 90 percent decrease in soil erosion, and from four to seven gallons (fifteen to twenty-six liters) of fuel saved per acre (Carpenter et al. 2002).

Unnecessary Speed Bumps and Roadblocks

Although the complexity of biological systems means that some promised benefits of biotechnology are many years away, the biggest threat that hungry populations currently face are restrictive policies stemming from unwarranted public fears. Although most Americans appear to be, at worst, indifferent to agricultural biotechnology, many Europeans and Asians have been far more cautious (Gaskell et al. 2000). Antibiotechnology campaigners in both industrialized and less-developed nations feed this ambivalence with scary stories that have led to the adoption of restrictive policies. Those fears are simply not supported by the scores of peer-reviewed scientific reports on the safety of biotechnology or the data from tens of thousands of individual field trials.

The evidence clearly shows no difference between the practices necessary to ensure the safety of bioengineered versus those needed for conventional plants (Institute of Food Technologists 2000). In fact, because more is known about the genes that are moved in transgenic breeding methods, it is actually easier to ensure the safety of bioengineered plants. But the public's reticence about transgenic plants has resulted in extensive regulations that require literally thousands of individual safety tests, many of them duplicative and largely unnecessary for ensuring environmental protection or consumer safety. In the end, overly cautious rules result in hyperinflated research and development costs and make it harder for poorer countries to share in the benefits of biotechnology.

While most surveys and focus-group reports show that anxiety about bioengineered food is at or near the bottom of almost every respondent's list of food worries, the issue of "public acceptance" has simply been made a scapegoat by politicians in many countries to slow down the adoption of food biotechnology. A recent commentary in the *Times* of London put it aptly:

> Asking people whether they're for or against GM crops is as ridiculous as asking whether they're for or against fire. As Prometheus found out, a mastery of flame can be a boon or a curse. It is the tool of the arsonist and [of celebrated chef] Gordon Ramsay. The technology is morally neutral. It is how it is applied that counts. (Henderson 2003)

Advocates have to take the public's concerns more seriously. Better sharing of information and a more forthright public dialogue are necessary to explain why scientists are confident that transgenic crops are safe. No one argues that we should not proceed with caution, but needless restrictions on agricultural biotechnology could dramatically slow the pace of progress and keep important advances out of the hands of people who need them. This is the tragic side effect of unwarranted concern.

The Road Ahead: Toward Improved Food Security for All

During the next fifty years, global population is expected to rise by 50 percent to 9 billion people, with nearly all of that growth coming in the poorest regions of the world (United Nations Secretariat 1999). Fortunately, humankind will face the extraordinary challenge of hunger and poverty with the very powerful tool of crop biotechnology. As noted, among the many causes of hunger during this century have been political unrest and corrupt governments, poor transportation and infrastructure, and, of course, poverty. All of these and more will need to be addressed if we are truly to conquer worldwide hunger. But ensuring true food security in a world of eight or nine billion will require greater food productivity.

As population increases, farmers must be able to grow more nutritious food on less land. Biotechnology can provide one very powerful way to do just that. Without such gains in productivity and nutrition, the growing need for food will require plowing under millions of hectares of wilderness—an environmental tragedy surely worse than any imagined by biotechnology's opponents. Furthermore, 650 million of the world's poorest people live in rural areas where agriculture is the primary economic activity. They are highly dependent upon the income that comes from growing and selling crops, so boosting the productivity of their crops would make a tremendous contribution to the battle against hunger and poverty.

Ultimately, biotechnology is more than just a new and useful agricultural tool. It could also be a hugely important instrument of economic development in many poorer regions of the globe. By making agriculture more productive, labor and resources could be freed for use in other areas of economic growth in nations where farming currently occupies 70 or 80 percent of the population. This, in turn, would be an important step in the journey toward genuine food security.

References

Adamu, Hassan. 2000. We'll Feed Our People as We See Fit. *Washington Post.* September 11, A23.

AgBioWorld Foundation. 2005. Scientists in Support of Agricultural Biotechnology. http://www.agbioworld.org/declaration/declaration_index.html (accessed February 14, 2005).

AgNews (Texas A&M University). 2003. Revolutionary Crop Yields Top List of Key Agricultural Events during Last 50 Years. April 3. http://agnews.tamu.edu/dailynews/stories/AGPR/Apr0303a.htm (accessed February 16, 2005).

Altieri, Miguel A., and Peter Rossett. 1999. Ten Reasons Why Biotechnology Will Not Ensure Food Security, Protect the Environment and Reduce Poverty in the Developing World. *AgBioForum* 2, no. 3–4 (Summer–Fall): 155–62.

American Medical Association. 2001. Genetic Enhancement Guards against Food Allergies. *AMA Media Advisory.* October 4.

Blackburn, Peter. 2003. Paraguay Farmers Opt for Banned GM Soybean Seeds. Reuters News Service. January 15.

Borlaug, Norman. 2000. Ending World Hunger: The Promise of Biotechnology and the Threat of Antiscience Zealotry. *Plant Physiology* 124 (October): 487–90.

Bright, Christopher. 1999. Invasive Species: Pathogens of Globalization. *Foreign Policy* 116 (Fall): 50–60.

Bruening, George. 2002. Safety of Foods Derived from Spliced-DNA Crops. Chapter 2 in *Benefits and Risks of Food Biotechnology*, ed. California Council on Science and Technology. Riverside, Calif.: California Council on Science and Technology.

Bruinsma, Jelle, ed. 2003. *World Agriculture towards 2030.* Rome: Food and Agricultural Organization.

Carpenter, Janet, Allan Felsot, Timothy Goode, Michael Hammig, David Onstad, and Sujatha Sankula. 2002. *Comparative Environmental Impacts of Biotechnology-Derived and Traditional Soybean, Corn and Cotton Crops.* Ames, Iowa: Council on Agricultural Science and Technology.

Commission of the European Communities. Directorate-General for Agriculture. 2000. Economic Impacts of Genetically Modified Crops on the Agri-Food Sector: A First Review. Working document, revision 2. Brussels: Commission of the European Communities.

Conko, Gregory. 2003. The Boons of Biotechnology. Chapter 6 in *Ecological Agrarian: Agriculture's First Evolution in 10,000 Years*, ed. J. Bishop Grewell and Clay J. Landry. West Lafayette, Ind.: Purdue University Press.

Conko, Gregory, and C. S. Prakash. 2002. The Attack on Plant Biotechnology. Chapter 7 in *Global Warming and Other Eco-Myths*, ed. Ronald Bailey. Roseville, Calif.: Prima Publishing.

Conway, Gordon, and Gary Toennissen. 1999. Feeding the World in the Twenty-First Century. *Nature* 402 (December 2): C55–58.

Della Penna, Dean. 1999. Nutritional Genomics: Manipulating Plant Micronutrients to Improve Human Health. *Science* 285 (July 16): 375–79.

Falck-Zepeda, Jose Benjamin, Greg Traxler, and Robert G. Nelson. 2000. Surplus Distribution from the Introduction of a Biotechnology Innovation. *The American Journal of Agricultural Economics* 82:360–69.

Gaskell, George, Nick Allum, Martin Bauer, John Durant, Agnes Allansdottir, Heinz Bonfadelli, Daniel Boy, Suzanne de Cheveigné, Björn Fjaestad, Jan M. Gutteling, Juergen Hampel, Erling Jelsøe, Jorge Correia Jesuino, Matthias Kohring, Nicole Kronberger, Cees Midden, Torben Hviid Nielsen, Andrzej Przestalski, Timo Rusanen, George Sakellaris, Helge Torgersen, Tomasz Twardowski, and Wolfgang Wagner. 2000. Biotechnology and the European Public. *Nature Biotechnology* 18:935–38.

Giannessi, Leonard. 2002. *Plant Biotechnology: Current and Potential Impact for Improving Pest Management in U.S. Agriculture: An Analysis of 40 Case Studies*. Washington, D.C.: National Center for Food and Agricultural Policy.

Goklany, Indur M. 1999. Meeting Global Food Needs: The Environmental Trade-Offs between Increasing Land Conversion and Land Productivity. *Technology* 6, no. 2–3:107–30.

Goodman, Robert M., Holly Hauptli, Anne Crossway, and Vic C. Knauf. 1987. Gene Transfer in Crop Improvement. *Science* 236 (April 3): 48–54.

Henderson, Mark. 2003. Who Cares What "the People" Think of GM Foods? *Times* (London). June 13.

Huang, Jikun, Scott Rozelle, Carl Pray, and Qinfang Wang. 2002. Plant Biotechnology in China. *Science* 295 (January 21): 674–77.

Institute of Food Technologists. 2000. *IFT Expert Report on Biotechnology and Foods*. Chicago: Institute of Food Technologists.

International Food Policy Research Institute. 2001. *The Unfinished Agenda: Perspectives on Overcoming Hunger, Poverty, and Environmental Degradation*, ed. Per Pinstrup-Andersen and Rajul Pandya-Lorch. Washington, D.C.: International Food Policy Research Institute.

Irish Examiner. 2003. GM Potatoes Set to Be Approved in India. June 11.

Ismael, Yousouf, Richard Bennett, and Stephen Morse. 2001. Farm Level Impact of Bt Cotton in South Africa. *Biotechnology and Development Monitor* 48 (December): 15–19.

James, Clive. 2004. *Global Status of Commercialized Transgenic Crops: 2004*. ISAAA Briefs No. 32. Ithaca, N.Y.: International Service for the Acquisition of Agri-Biotech Applications.

Kendall, Henry W., Roger Beachy, Thomas Eisner, Fred Gould, Robert Herdt, Peter Raven, Jozef S. Schell, and M. S. Swaminathan. 1997. *Bioengineering of*

Crops: Report of the World Bank Panel on Transgenic Crops. Washington, D.C.: World Bank.

Kessler, Charles, and Ioannis Economidis, eds. 2001. *EC-Sponsored Research on Safety of Genetically Modified Organisms: A Review of Results*. Luxembourg: Office for Official Publications of the European Communities.

Kirsten, Johann, and Marnus Gouse. 2002. Bt Cotton in South Africa: Adoption and Impact on Farm Incomes amongst Small and Large-Scale Farmers. *ISB News Report*. October, 7–9.

Langridge, William H. R. 2000. Edible Vaccines. *Scientific American*, September.

Maluszynski, M., K. Nichterlein, L. Van Santen, and B. S. Ahloowalia. 2000. Officially Released Mutant Varieties: The FAO/IAEA Database. Vienna, Austria: Joint FAO-IAEA Division, International Atomic Energy Agency. December.

Manuel de la Fuente, Juan, Verenice Ramírez-Rodríguez, José Luis Cabrera-Ponce, and Luis Herrera-Estrella. 1997. Aluminum Tolerance in Transgenic Plants by Alteration of Citrate Synthesis. *Science* 276 (June 6): 1566–68.

Mgendi, Catherine. 1999. Local Scientists Snub the West in Biotech War. Africa News Service. October 21.

National Academy of Sciences. 1987. *Introduction of Recombinant DNA-Engineered Organisms into the Environment: Key Issues*. Washington, D.C.: National Academy Press.

Pew Initiative on Food and Biotechnology. 2001. *Harvest on the Horizon: Future Uses of Agricultural Biotechnology*. Washington, D.C.: Pew Initiative on Food and Biotechnology. September.

Prakash, C. S. 2001. The Genetically Modified Crop Debate in the Context of Agricultural Evolution. *Plant Physiology* 126 (May): 8–15.

Qaim, Matin, and David Zilberman. 2003. Yield Effects of Genetically Modified Crops in Developing Countries. *Science* 299:900–902.

Rissler, Jane, and Margaret Mellon. 1996. *The Ecological Risks of Engineered Crops*. Cambridge, Mass.: MIT Press.

Rockefeller Foundation. 1999. New Rices May Help Address Vitamin A and Iron Deficiency, Major Causes of Death in the Developing World. Press release, August 3.

Royal Society of London, the U.S. National Academy of Sciences, the Brazilian Academy of Sciences, the Chinese Academy of Sciences, the Indian National Science Academy, the Mexican Academy of Sciences, and the Third World Academy of Sciences. 2000. *Transgenic Plants and World Agriculture*. Washington, D.C.: National Academy Press.

Sommerville, Chris, and John Briscoe. 2001. Genetic Engineering and Water. *Science* 292 (June 22): 2217.

United Nations Development Programme. 2001. *Human Development Report 2001: Making New Technologies Work for Human Development*. New York: Oxford University Press.

United Nations Secretariat. Department of Economic and Social Affairs. Population Division. 1999. *The World at Six Billion*. Working paper ESA/P/WP 154. New York: United Nations.

Wambugu, Florence. 2001. *Modifying Africa*. Nairobi, Kenya: Florence Wambugu.

White, Ed. 2001. GM Canola: Who Benefits? *Western Producer.* July 5.

Ye, Xudong, Salim Al-Babili, Andreas Klöti, Jing Zhang, Paola Luca, Peter Beyer, and Ingo Potrykus. 2000. Engineering the Provitamin A (ß-Carotene) Biosynthetic Pathway into (Carotenoid-Free) Rice Endosperm. *Science* 287 (January 14): 303–8.

3

Trade War or Culture War?
The GM Debate in Britain and the European Union

Tony Gilland

The European Union's approach to assessing the implications of genetically modified (GM) crops and food for human health and the environment has been based on shaky scientific foundations from the start. Although the EU ended its de facto moratorium on GM crops in the latter half of 2004 (it had been in effect since 1998), five member states remain defiant and are maintaining national bans. The acrimony and the World Trade Organization (WTO) case between the United States and the EU continue.

Commentators trying to make sense of the European position have argued that the EU and various member states are trying to protect their oversubsidized farmers from the pressures of international competition, and that opposition to GM is part of this anti-free-trade strategy. While there might well be important trade aspects to this rift between Europe and the United States, the issues at stake are more profound than can be adequately explained in terms of a trade war. Rather, we appear to be witnessing a culture war about the sort of society we should want to live in, which often takes the form of a battle over methods of food production—GM versus conventional agriculture, organic versus intensified—and recently has appeared as a conflict between Europe and America.

Europe and the Precautionary Principle

The EU and its member states are most often guided on issues of science and policy by the precautionary principle—the assumption that experimentation should only proceed when there is a guarantee that the outcome will not be harmful. In *The Journal of Risk Research*, Jonathan Wiener and Michael Rogers argue that many proponents of the precautionary principle regard it as "an antidote to industrialization, globalization and Americanization" and contrast a "civilized, careful Europe" with the "risky, reckless and violent America" it confronts—a view that demonstrates the positive cultural meanings some Europeans attach to the precautionary principle. Others, they point out, interpret the principle differently—evincing, for example, the image of "a statist, technophobic, protectionist Europe trying to rise to challenge a market-based, scientific, entrepreneurial America" (Wiener and Rogers 2002, 334).

However, the main thrust of Weiner and Rogers's paper is to challenge the commonly held assumption that "Europe has become more precautionary than the U.S. since the 1990s." While they acknowledge that the European Commission has formally articulated and endorsed the precautionary principle, and that this is an important distinction between the EU and the United States, they show how the U.S. government and institutions have been more than happy to apply the principle in different policy disputes. According to their analysis, Europe appears to be more precautionary than the U.S. when it comes to "GMOs [genetically modified organisms], hormones in beef, toxic substances, phthalates, climate change, guns, and antitrust/competition policy," while U.S. actions appear more precautionary in relation to a whole range of other issues, from new drug approvals to lead in gasoline, particulate air pollution, cigarette smoking, and restrictions on imports of beef (Commission of the European Communities 2000).

Regardless of whether we accept this analysis, it is an important reminder that a heightened preoccupation with risk has become a global phenomenon and is by no means confined to specific issues such as GM food. Fierce debates about global warming, biodiversity, waste, nuclear power, sustainable development, electromagnetic fields, human genetics, and, more recently, nanotechnology play out across the world. The specific issues do not spur these debates. Rather, they are part of a broader discussion about the way we view the world and the kind of future society we wish to create. Alongside every discussion of GM food, global warming, human genetics, or new medicines lie fundamental questions about the role of science and technology, man's relationship with nature, democracy, and the role of corporations.

It is in this context that we should view the issue of GM crops and food. Operating within the climate of the precautionary principle, politicians, scientists, and industrialists favorable toward GM technology have found themselves poorly equipped to make a positive case for its implementation. Hypothetical worst-case scenarios, often with little theoretical plausibility, have been highlighted alongside the charge that there is little benefit to consumers from this technology—so why should they accept even the most minimal of risks? Too often, the response of authority figures to these challenges has been to put in place increasingly restrictive regulatory controls and to commission yet more research, in an attempt to assuage the unassuageable demand that there will be no unforeseen adverse consequences from the application of this technology.

The bumbling way in which policymakers have handled the issue of GM crops and food, particularly within Europe, has established worrisome precedents for the way modern societies relate to science, technology, and innovation. How did we get into this situation? It is worth reviewing how Europe's attitudes toward GM technology have unfolded.

The British Experience

Technical and regulatory responses to public concerns about health and environmental issues have not worked in Europe. Of course, reasonable regulations should be employed when genuine issues of safety or adverse consequences are at stake. But this is not what the EU's regulatory approach to GM crops and food has been about. Working first of all in anticipation of potentially negative public reactions to GM, then in response to campaign group and media pressure, and next in response to the position adopted by the food retail industry, the EU has introduced increasingly tight regulatory controls of GM technology.

The clear aim of these controls has been to address questions of public perception, rather than scientific questions of health or environmental safety. Consequently, the regulations inevitably end up embodying exaggerated, if not spurious, concerns in an attempt to reassure the public that everything is being done to protect them. Rather than being properly addressed, challenged, and put into perspective, these concerns are then validated and institutionalized. And once governments and regulatory bodies have demonstrated their willingness to regulate for the hypothetical, the door is opened to further demands for regulation based on equally hypothetical concerns.

The earlier period in the history of Europe's path toward the commercialization of GM crops, from the late 1980s to the mid-1990s, is noteworthy to the extent that it was primarily a period of anticipating public concerns and regulating to head them off. The key plank of EU regulation governing the commercial release of GMOs—EC Directive 90/220, implemented in 1991—embodies the assumption that the process of genetic modification itself is inherently suspicious. Mark Cantley is a senior researcher with the European Commission Research Directorate and former head of the biotechnology unit in the Directorate for Science, Technology and Industry of the Organisation for Economic Co-operation and Development (OECD). He believes that political concerns and fears about reactions to GM technology led the EU in 1986 to ignore the scientific judgment of the OECD (validated by the experience of the subsequent seventeen years) that "there is no scientific basis for specific legislation to regulate the use of recombinant DNA organisms" (Cantley 1995).

In a similar vein, though against a backdrop of greater controversy surrounding the GM issue, the EU Novel Foods Regulation in May 1997 introduced mandatory consumer labeling requirements for foods containing GM ingredients. Given that GM foods approved for marketing are judged safe, there is no scientific health information being communicated to consumers under this provision.

The British experience of the GM debate in the late 1990s provides a useful case study of how the "public concerns" that have so influenced EU regulation have been shaped. In December 1997, the United Kingdom's influential left-leaning national broadsheet, the *Guardian*, embarked upon a four-day campaign to bring the potential perils of GM food to the nation's attention. The paper was building on the work of others such as Greenpeace

UK, which was in search of a new issue following the success of its Brent Spar campaign against the Shell Oil Company, and the Consumers' Association, an influential UK consumer lobby group primarily concerned with the question of consumer choice and the labeling of GM foods.[1]

On the first day of its series on GM food, the *Guardian's* front-page story, "Food: The £250 Billion Gamble—Big Firms Rush for Profits and Power Despite Warnings," warned of "millions of farmers unemployed," "poor countries losing whole export markets," and "consumer revolt in Europe." It pointed out that "the first commercial releases of genetically engineered seeds are expected to be approved by the European Union early next year." The paper promised a special analysis of "unexpected environmental problems," "consumers being given no effective choice over foods," "heavy lobbying to rewrite world food safety standards in favour of biotechnology," and an apparent "revolving door between the U.S. government and the biotech industry."

The next day, the *Guardian* reported that the UK government had decided to delay issuing licenses for GM crops, and expressed its fear that Whitehall had "underestimated the dangers of the new food revolution." Interestingly, only a few weeks earlier, Jeff Rooker, then minister for food safety, had appeared on BBC television to state that he was satisfied with the regulatory controls in place to deal with GM foods, and comfortable with the decision to grant a license for a number of novel crops to be grown in the United Kingdom starting in 1998 (Rooker 1997). The New Labour government's impulse to modify its position to appease the media was indicative of how it mishandled the GM debate.

By the end of 1998, things looked far worse for the commercial application of GM crops, both in the United Kingdom and elsewhere in Europe. In the United Kingdom, a minor but well-known frozen food retailer, Iceland, announced that it had managed to source GM-free soya, and guaranteed that all its own-brand products would no longer contain any GM ingredients. The chairman of Iceland, Malcolm Walker, was moved to remark, "The introduction of genetically modified ingredients is probably the most significant and potentially dangerous development in food production this century." Walker collaborated with Friends of the Earth and Greenpeace to promote an opportunistic "disloyalty card," targeting the Unilever food manufacturing company for supposedly contaminating its brand-name products (brands that Iceland stocked and sold) with GM ingredients. Prince Charles published a high-profile article in the well-respected *Daily Telegraph*, decrying GM crops for "taking mankind into realms that belong to God, and to God alone." This stand received much praise from campaign groups and commentators who regarded themselves as radical and antiestablishment (*Daily Telegraph* 1998).

It was at this point that English Nature, the government's statutory conservation body, called for a three-year moratorium on the commercial growth of GM crops, expressing concern that they would lead to the further intensification of farming with negative consequences for farmland wildlife. Much sympathetic coverage was given to organic farmers protesting that GM crops would "pollute" their products through cross-pollination, and to campaigners ripping up scientific crop trial sites. In August 1998, Arpad Pusztai, a researcher at the Rowett Institute in Scotland examining the effects of feeding a variety of GM potato to rats, made headline news with his apparent discovery of potentially negative effects on the rats' growth and immune systems. Pusztai told the *World in Action* documentary program that he thought it "very unfair to use our fellow citizens as guinea pigs." As was well known at the time, Pusztai's research had not been peer-reviewed. A subsequent panel of reviewers from the Royal Society—the United Kingdom's preeminent body of scientists—found this piece of research to be "flawed in many aspects of design, execution and analysis," and warned that "no conclusions should be drawn from it" (Royal Society 1999).

An unusual alliance of groups and organizations, claiming to represent over four million individuals, promoted a "Five Year Freeze" on the implementation of GM technology (Five Year Freeze Campaign 1998). These organizations ranged from high-profile environmental and consumer groups and aid agencies to the less obvious Townswomen's Guild, the major trade unions, UNISON and TGWU, and the Local Government Association. At this time, it almost seemed as though every organization wanted to jump on the anti-GM bandwagon. (Interestingly, the campaign's current Web site provides a hit counter showing that it received 4,509 visits in the 354-day period following March 11, 2004. Clearly, the organization has not maintained any lasting and substantial grass-roots appeal, despite its proclivity for speaking on behalf of millions of the public.) With continuing contrversy over the labeling of foods containing GM ingredients and confusion as to the interpretation of EU regulations, the issue of "consumer choice" remained at the top of the agenda. Environmental groups in particular used the issue of labeling to good emotive effect: Having first scared consumers about the unknown ill-health effects of GM, they set about portraying biotechnology companies (particularly Monsanto), food producers and retailers, and governments as "force-feeding" consumers an inherently suspicious product.

Media hysteria rather than public hysteria finally sealed the near-term fate of GM commercialization in the United Kingdom. A petition crafted by campaigners and signed by twenty scientists, primarily objecting to the dismissal of Arpad Pusztai from his job at the Rowett Institute rather than defending his discredited research results, sparked a ten-day media frenzy covering all aspects of GM, beginning in February 1999. Within a short time, supermarkets had distanced themselves from GM products and even began competing with one another to prove to their customers that their own brands would no longer contain GM ingredients. Tesco, the largest supermarket in the United Kingdom, followed Asda, Sainsbury, and Waitrose, announcing at the end of April 1999 that it would "remove GM ingredients where we can and label where we can't" (*Independent* 1999).

Unilever announced at the same time that it would remove GM ingredients from its products in response to consumer concerns. A spokesman for Van den Bergh, a subsidiary of Unilever, cited "a sharp rise in calls to consumer helplines" as demonstrating "a significant shift in consumer perception after an unprecedented media campaign" and as justifying the company's decision—despite the fact that, throughout the campaign against GM foods the previous year, "sales had held up well" (*Times* [London] 1999a). Other companies similarly cited a sharp rise in consumer calls as the reason for their having abandoned GM ingredients. Alison Austen, a senior manager at Sainsbury's Supermarkets Ltd., told an Institute of Grocery Distribution conference in July 1999 that the company received 10,500 inquiries to its consumer care line in February 1999 (Institute of Grocery Distribution 1999). Though obviously a large number of phone calls, they were clearly in response to the unprecedented, sustained, and alarmist media coverage that month; and, as Austen told the conference, the calls decreased by two-thirds to 3,000 the following month, and continued dropping rapidly after that. Furthermore, the fact that Sainsbury's had millions of customers and a market share of over 15 percent in the United Kingdom should put these figures into some perspective. Lord Haskins, chairman of the major British food manufacturer Northern Foods, and of the government's Better Regulation Task Force, told the *Times* in June 1999 that he was "ashamed at the way retailers have wobbled"—although he had to admit that his own company had decided to stop using GM ingredients.

Food Scares and the Significance of the BSE and CJD Episode

There is a widely held view that the mishandling by the UK government and regulatory authorities of the widespread appearance of bovine spongiform encephalopathy (BSE) in cows in the late 1980s, and the subsequent identification of a new variant of the fatal brain disease in humans, Creutzfeldt-Jakob disease (CJD), explain the reaction against the introduction of GM crops and food. However, it is important to remember that the so-called mad cow panic occurred at a time when risk-averse trends and obsessions with worst-case scenarios were already deeply embedded within British and European societies, and that this cultural climate had a major impact on the interpretation of the bovine spongiform encephalopathy and Creutzfeldt-Jakob disease episode.

In the late 1980s, environmental issues rapidly moved up the political agenda. As British sociologists Phil Macnaghten and John Urry describe in their book, *Contested Natures*, Conservative members of Parliament began to notice that environmental issues had some popular resonance, and they incorporated them into their speeches and campaigns (Macnaghten and Urry 1998). In September 1988, in the face of declining political influence, Margaret Thatcher, who had only a few years earlier been dismissive of concerns over such issues, gave a headline-grabbing speech at the Royal Society about the "unwitting experiment" that humanity had been conducting with the planet. The media responded positively, which helped trigger a massive expansion in the memberships of Greenpeace (UK membership increased from 50,000 to 320,000 between 1985 and 1989) and Friends of the Earth (from 27,000 to 120,000), bringing the collective membership of the British environmental movement to 4.5 million by 1990. By the time of the Rio Earth Summit of 1992, the concept of sustainability, with its embrace of the notion that there are natural limits to human activity, was becoming a necessary and important part of corporate-speak.

The first major food scare in the United Kingdom occurred in 1989, when the then health minister Edwina Currie went on television to announce that many eggs and chickens contained salmonella. Hysteria followed, egg sales plummeted, and costly salmonella programs were implemented. Interestingly, at the time, the announcement by Currie was seen as irresponsible and alarmist in political circles, and she promptly resigned in disgrace. As Alan McHughen notes in his book *A Consumer's Guide to GM Food*, the incidence of salmonella at the time was about 1 in 650 eggs; today, after implementation of the costly control programs, the incidence is at much the same level (McHughen 2000). Seen in this light, Currie's resignation seems fair enough. But in recent years, during an inquiry into BSE, Currie was praised retrospectively for her courage in acting in a precautionary way and for lifting the lid on the complacent culture at the Ministry of Agriculture, Fisheries and Food (now the Department for Environment, Food and Rural Affairs). This shows a striking shift in political culture over less than a decade.

The salmonella scare was followed by a number of other minor scares about food safety. The twin concerns of environmental catastrophe and new dangers to human health began to have an increasingly dominant influence over the public mood; risk-aversion became the order of the day.

Bovine spongiform encephalopathy is a prion disease—that is, a disease that primarily affects the nervous system. First discovered in British cattle in 1986, it was quickly found to have resulted from feeding sheep offal to cows to boost their protein intake. The practice was banned in 1988, and measures were introduced to prevent the transmission of BSE to humans. Based on the available evidence, scientists at the time thought it highly unlikely that the disease would cross the species barrier to humans.

In 1995, ten cases of a distinctive variant of the fatal brain disease Creutzfeldt-Jakob disease, also a prion disease like BSE and scrapie in sheep, were identified. This new variant of the disease was subsequently referred to as new variant CJD, or nvCJD. Casualties of the disease had a younger age profile than usual, and there were distinctive clinical and pathological manifestations. The advisory committee established by the government to monitor the BSE situation and advise on human health implications changed its position, from claiming there was no evidence that the disease could infect humans to claiming that a link between BSE and

CJD was the most likely explanation for the ten cases—despite the fact that no causal mechanism for the transfer of the disease had been identified.

Predictions of hundreds of thousands, and in some cases millions, of deaths abounded in the media, while politicians and the Chief Medical Office did little to quell such fears and in many ways endorsed them. For example, the CMO faxed medical doctors, alerting them to this new interpretation of the situation. This measure had little public health value, since stringent measures to protect the public had already been put into place. By May 2003, only 129 cases of nvCJD had been identified (96 confirmed and the remainder probable), and the incidence of CJD may now have reached its peak.

The point is not to dismiss people's fears about BSE and CJD, but rather to emphasize how public anxiety fueled an alarmist response to a genuine concern, which then had a negative impact on people's perceptions of food safety, science, and government regulation. An unfortunate biological fluke, speedily responded to once identified, was turned into a grim moral story for our times about the dangers of human greed, government coverup, and nature's revenge.

But the domestication of animals has been a source of infectious disease since Neolithic man. Measles, mumps, whooping cough, smallpox, and tuberculosis all crossed the species barrier at some stage, with catastrophic consequences. In a different political and cultural climate, the BSE/CJD episode might be read as an example of how much better society has become at dealing with these problems (Fitzpatrick 2001).

The UK Government's Response to GM Fears

The first labeled GM food product reached the United Kingdom at the height of the mad cow hysteria. At the time, one unlabeled GM food, "vegetarian" cheese made with an enzyme engineered into a microbe rather than extracted from the lining of calf stomachs, was being sold with no controversy. Some vitamins and other food supplements came from fermentation processes using genetically modified organisms, and a number of medical drugs, including human insulin, were made by similar processes. In 1996, tomato paste made from GM tomatoes with a mutation to allow ripening without softening appeared on the shelves of two grocery chains. It was clearly labeled "*Made with genetically modified tomatoes*," and was shelved side by side with a similar product made from conventional tomatoes. Both cans sold for the same price, but the GM one was larger at 170 grams, compared with 140 grams, thus offering a price saving of some 20 percent. Sales of both types of can were roughly equal.

But by the second half of 1999, all of the major supermarket chains in the United Kingdom had responded to public pressure by withdrawing GM products—not only the tomato paste, but also GM herbicide-tolerant or insect-resistant soya and corn products. Restaurants were now obliged to indicate whether their dishes contained GM products, and many withdrew GM foods from their menus. How could such a radical change in the public's response come about within three years?

Throughout the unfolding controversy surrounding GM crops in the United Kingdom, it is hard to identify a statement by a government minister that did not encourage increased regulation and precaution. The heightened sensitivity toward the destructive aspects of human innovation that risk aversion represents shaped the Labour government's response to GM technology. Though ministers emphasized that there was no evidence GM foods were unsafe for humans to eat, their overriding message was hardly reassuring.

In May 1998, the then agriculture minister Jack Cunningham announced that following hard campaigning by the British government, the EU had agreed to new rules on GM labeling that would require "all foods containing ingredients produced from Monsanto's GM soya and Novartis' GM maize to be labeled except when neither protein nor DNA resulting from genetic modification is present." Though he pointed out that GM could offer cheaper and more nutritious foods for consumers, his primary goal was clearly to align the government with perceived consumer concerns, by boasting of the government's role in bringing about this supposedly important change. After arguing that the government's role should be to help ensure that consumers could "choose whether to eat genetically modified foods or not," he also announced that the government was "looking at ways of further strengthening" its "monitoring and testing procedures" to protect human health, wildlife, and the environment (United Kingdom, Ministry of Agriculture, Fisheries, and Food 1998a).

In October 1998, in widely publicized statements to a House of Lords Select Committee, Food Safety Minister Jeff Rooker and Environment Minister Michael Meacher announced that the government was seeking yet further tests and controls over genetically modified organisms (GMOs). These included the seeking of an amendment to EC Directive 90/220 to expand its scope to "cover indirect as well as direct effects of GMOs" (United Kingdom, Ministry of Agriculture, Fisheries, and Food 1998b). Meacher's statement sought to gain public favor by presenting the British government as arguing for even greater caution than the EU. He portrayed the interests of consumers and the public as being in conflict with those of industry, stating that "our aim is to strike the right balance between protecting the environment and human health on the one hand, and on the other, maintaining the proper degree of certainty needed by business for the development of new products." Rooker took the extraordinary step of telling the committee that, although GM foods were rigorously assessed for human safety, the government was "currently looking into the possibility of going even further by introducing monitoring arrangements capable of picking up any unexpected effects should they emerge in the future" (United Kingdom, House of Lords 1998).

It is hard to see how people can be reassured about a technology when there is an official emphasis upon the need for existing rigorous controls to "go further." This was characteristic of the UK government's response. Each time controversy hit the headlines, the government introduced ever-tighter regulations to placate the critics and respond to perceived public anxieties. This announcement about monitoring the long-term health effects of GM foods had nothing to do with science, since scientists have no idea of how to conduct such a task meaningfully. Rather, it was an opportunistic and unhelpful concession to the notion that there might just be something scary about this technology. Scientific advisory committees to the government demonstrated a similar crisis of confidence. All too often, they made pronouncements that pandered to the public mood of danger and uncertainty, rather than deploying the scientific evidence to address exaggerated concerns.

An interesting example of the impact of perceptions on scientific advice was the recommendation of the government's Advisory Committee on Novel Foods and Processes (ACNFP) in 1996, prior even to the high-profile campaign against GM foods. The committee argued that Novartis's Bt maize should not be granted regulatory approval because the modified crop contained a marker gene resistant to the antibiotic penicillin. Discussing this decision at a later inquiry into GM regulations, the chair of the committee, Professor Derek Burke, explained that the committee was aware that the chances of this product increasing antibiotic resistance within humans "were unlikely to be as much as one in a million million," but he justified the decision on the basis that the committee was "dealing more with people's perceptions and concerns" (United Kingdom, House of Lords 1998).

The use of antibiotic-resistant marker genes within the process of genetically modifying crops is an ongoing source of public controversy, despite the absence of any scientific evidence of a risk to human health. This is surely in part due to the reluctance of some official advisory committees to employ scientific evidence to address concerns for fear of courting disfavor. It is worth noting that this approach has not been restricted to the United Kingdom, or even Europe. In July 2000, for example, the Brazilian Academy of Sciences, the Chinese Academy of Sciences, the Indian National Science Academy, the Mexican Academy of Sciences, the National Academy of Sciences of the United States of America, the Royal Society (United Kingdom), and the Third World Academy of Sciences jointly published a report on the benefits and risks associated with GM crops. On this issue of antibiotic-resistant marker genes they stated:

> No definitive evidence exists that these antibiotic resistance genes cause harm to humans, but because of public concerns, all those involved in the development of transgenic plants should move quickly to eliminate these markers. Ultimately, no credible evidence from scientists or regulatory institutions will influence popular public opinion unless there is public confidence in the institutions and mechanisms that regulate such products (Brazilian Academy of Sciences et al. 2000).

The inference is clear: Unless regulations are based on public perceptions, no matter how exaggerated, there will be a lack of public confidence in regulatory institutions. What the authors of this report failed to recognize is that, if scientific evidence is silenced in favor of pure public relations, then the basis of any scientific committee or regulatory body is undermined, and becomes arbitrary.

Unfortunately, this is precisely the sort of approach adopted by the United Kingdom, and it has yet to yield any constructive results. In the summer of 1999, the government announced an overhaul of the advisory system on GM designed to make it clear to the public that perceptions of danger would play a more prominent role in regulation and policymaking decisions concerning the future of GM technology. The government said that the new advisory system would include a broader range of interests—"Those with expertise of consumer issues and ethics, for example, will sit alongside scientists." It also restated its intention to consider "the establishment of a national surveillance unit to monitor population health aspects of genetically modified and other types of novel foods" (United Kingdom, Cabinet Office 1999). By the end of 1999, the government had secured the biotechnology industries' agreement to a three-year voluntary moratorium on the commercial growth of GM crops in the United Kingdom, on top of the year-long moratorium it had negotiated with the industry at the end of 1998 (United Kingdom, Department of the Environment, Transport and the Regions 1999).

On February 27, 2000, Prime Minister Tony Blair, in a high-profile statement published in the *Independent on Sunday* and widely reported the following day, told the nation that "no GM crops will be grown commercially in this country until we are satisfied there will be no unacceptable impact on the environment." (He went on to contrast this position with the more cavalier approach of the United States, "where an area the size of Wales is already under cultivation"; Blair 2000.) The statement was widely reported as a U-turn by the British prime minister, who was well known to be favorable toward the use of GM technology. Like the supermarkets, Blair eventually bowed to the constant barrage of complaints coming from some quarters, demonstrating that his desire to be seen to be in tune with perceived public opinion was stronger than his will to provide strong political leadership.

In the UK debate about GM, a strong positive political motivation for the technology was conspicuously absent. Rather than promoting the importance to society of experimenting with new technology in general, and agricultural biotechnology in particular, the British government became fixated upon proving its credentials in relation to consumer safety and the environment. The new food technology became politicized to the extent that whatever science showed about the risks and benefits of GM, and however beneficial this technology might be from the standpoint of agricultural production and profits, anybody wanting to appear in line with public concerns was obliged to express precaution and determination to reduce potential risks.

European Regulation and Labeling Requirements

A similar process has been at work throughout Europe. In May 2003, in response to the news that the United States would be filing a WTO case against the EU over its anti-GMO position, EU Commissioner for Health and Consumer Protection David Byrne announced that Europe had "been working hard" to "complete our regulatory system in line with the latest scientific and international developments. . . . Unless consumers see that the authorization process is up to date and takes into account all legitimate concerns, consumers will continue to remain skeptical of GM products." EU Commissioner for the Environment Margot Wallstrom added, "The Commission strongly believes that we in Europe should move ahead with completing our legislation on traceability and labeling and on food and feed, currently before the European Parliament. We should not be deflected or distracted from pursuing the right policy for the EU" (Commission of the European Communities 2003).

The key phrase here is "the right policy for the EU." Byrne and Wallstrom argue that a scientific assessment of the health and environmental implications of GM crops and food is culturally specific to the EU member states, and that it therefore has to take account of the prevailing political climate. It is implied that by bowing to these cultural concerns, the EU is adopting a more responsible approach and ultimately stands a better chance of winning the public over. The EU now hopes that its proposed complex system of labeling and traceability requirements, which will require the labeling of any food produced from a GM organism regardless of the presence or absence of novel genetic material, will allow the moratorium to be lifted and the EU to defend itself against any public outcry. Byrne and Wallstrom criticized the timing of the U.S. filing of its WTO case as "eccentric," and threatening to the good work the EU had done in getting this far.

Yet, as recent history tells us, such an approach only exacerbates unfounded fears. The character of these regulatory requirements is informed by the unwillingness of those in authority to challenge the risk-averse mood of our times. Instead, a system of regulation is established that, first, reinforces the idea that something might go wrong and GM might indeed poison us; and second, tells the consumer that the ultimate decision over the safety of the food he purchases is his choice and responsibility.

Ultimately, the consumer will, indeed, resolve the labeling issue. But it is important to be clear about what is being said here. We are witnessing a major abdication of responsibility on the part of the EU and member-state governments. What is the point of having an army of scientific experts investigating every aspect of this technology if they are not allowed to give the benefit of their expertise? Either the EU, based on the advice of its scientists, believes that GM technology is generally safe and worth making use of, or it does not. But rather than tell the public what it really thinks and therefore take responsibility for making a decision, the EU would cover its back and allow consumers, who cannot expect to be well informed on the safety of GM food, to take that responsibility. Their message is, "Don't listen to us; make up your own minds." We have ended up in a situation where a state-backed system of labeling based on misconceptions rather than science is put forward as a positive example of consumer choice.

The current labeling regulations, which have equivocation about GM built into the procedure, have raised new problems. The primary focus of governments within Europe is now about how to prevent the contamination of conventional and organic food by cross-pollination with GM varieties. Many hours have been spent debating over what minute levels of contamination of conventional and organic produce by GM varieties should be deemed acceptable as still providing consumers with a non-GM product choice. The labeling of GM food products, originally conceived of as a way to placate the critics of the technology, has descended into complex, detailed, and farcical discussions about ensuring that consumers can avoid the tiniest quantities of GM ingredients even when consuming heavily processed foods.

While the EU, since late 2004, has begun to approve GM crops for cultivation again, its regulatory approach has clearly not won over hearts and minds. After all, it is difficult to hold out against the use of technology on irrational grounds indefinitely. Importantly, European farmers are unlikely to embrace GM crops in the way U.S. farmers have done for a long time to come. Through this whole saga, Europe has moved a long way down the route of defining its own culturally acceptable approach to the regulation of risk-related issues. Rather than confront the broader climate of fear, Europe has chosen to accommodate to it and in so doing has set some very negative and dangerous precedents.

The Price of the Precautionary Principle

There are numerous candidates for blame in the GM debacle. National governments have vacillated, the media have scaremongered, and campaign groups and NGOs have often exaggerated even the slightest concerns. Retailers, too, deserve some of the blame, for the speed with which they have dumped GM products whenever they have caught a whiff of controversy. However, to apportion blame in this way would be to gloss over the more profound processes at work in society that shaped the GM debate. The dangers go beyond the application of this particular technology.

The way in which the GM issue has been handled in the United Kingdom and Europe indicates just how much the precautionary principle has taken hold as an organizing principle of modern society. The scientific issue of the risks and benefits of GM crops is now subsumed beneath an overarching political concern to hold back lest something should go wrong. That a potentially beneficial technology such as this could be retarded "just in case" indicates a fundamental value shift within modern society that has practical consequences for how we develop science and technology. This is a shift from the belief that progress is a social good, and that science and technology should be developed for the benefit of humanity, to a distrust of the consequences of progress. At worst, it could result in a severe overreaction, restricting science and technology even when the potential rewards greatly eclipse the potential risks.

Why has this value shift taken place? It is not because life has become more dangerous. We live healthier, wealthier, and safer lives than at any time in history. It is also clear that reactions against certain new technologies have very little to do with the technologies themselves. Despite the fearful focus on GM crops and food in the United Kingdom and Europe, and the never-ending process of testing and monitoring this technology, there remains no scientific evidence that GM crops will actually harm humanity or the environment.

The resistance to agricultural biotechnology reflects a growing distrust of political authority and scientific expertise. Combined with an increasingly individuated consumer society, this has led to a situation in which unfounded fears can take hold very rapidly, spread by unofficial sources such as the media, campaign groups, and maverick scientists. Rather than attempting to counter these scares directly, the official approach has been to bend over backwards to take such fears, no matter how exaggerated, into account—thereby implicitly endorsing them. Thus, unfounded fears have given rise to unfounded regulation, which, in turn, creates the basis for yet more fear and more regulation.

Note

1. Greenpeace hit the front pages in June 1995 when its campaign to prevent the sinking of the Brent Spar oil platform led to a humiliating retreat by the Shell Oil Company. Greenpeace warned of the danger posed to "all marine and human life" by sea-dumping of platforms and rigs. Though this was a false claim—even Greenpeace later admitted to getting its facts wrong—the organization was widely applauded for challenging mankind's destructive behavior toward the natural environment. Shell and other oil companies such as BP promised to change their ways and to consult campaign groups like Greenpeace about how best to do so. In July 1997 Greenpeace published a report, *From BSE to Genetically Modified Organisms—Science, Uncertainty and the Precautionary Principle*, in which it sought to tap into consumer concerns about BSE to argue the case for precaution in relation to GM.

In 1997, the Consumers' Association published a report, "Gene Cuisine," which raised a variety of concerns about the unpredictable consequences of GM technology but was relatively mild-mannered compared to the subsequent alarmist campaigning that went on. The report did, however, make much out of the issue of consumer choice and labeling, and argued that the EU's regulations should be tightened to ensure that "all foods or food ingredients derived by the use of genetic modification should be labelled" including "food and food ingredients where the genetically modified material is no longer intact or 'live.'"

References

Blair, Tony. 2000. The Key to GM Is Its Potential, Both for Harm and Good. *Independent on Sunday*. February 27.

Brazilian Academy of Sciences, Chinese Academy of Sciences, Indian National Science Academy, Mexican Academy of Sciences, National Academy of Sciences of the USA, the Royal Society (UK), and the Third World Academy of Sciences. 2000. *Transgenic Plants and World Agriculture*. Washington, D.C.: National Academy Press.

British Department of Health Monthly. 2003. Creutzfeldt-Jakob Disease Statistics. May 6.

Cantley, Mark. 1995. The Regulation of Modern Biotechnology: A Historical and European Perspective. In vol. 12 of *Multi-Volume Comprehensive Treatise: Biotechnology*, 2nd ed., ed. H. J. Rehm and G. Reed. Weinheim, Germany: VCH.

Commission of the European Communities. 2000. Communication from the Commission on the Precautionary Principle. EC COM 1 final. Brussels: Commission of the European Communities. February 2.

———. 2003. European Commission Regrets U.S. Decision to File WTO Case on GMOs as Misguided and Unnecessary. Press release. Brussels: Commission of the European Communities. May 13.

Consumers' Association. 1997. *Gene Cuisine*. Report published by the Consumers' Association.

Daily Telegraph. 1998. Seeds of Disaster. June 8.

Fitzpatrick, Michael. 2001. *The Tyranny of Health*. London: Routledge.

Five Year Freeze Campaign. 1998. http://www.fiveyearfreeze.org. July.

Greenpeace. 1997. From BSE [bovine spongiform encephalopathy] to Genetically Modified Organisms—Science, Uncertainty and the Precautionary Principle. Report published by Greenpeace. July.

Independent. 1999. Tesco and Unilever Ban GM Products. April 28.

Institute of Grocery Distribution. 1999. Author's notes taken at the IGD "Consumer Trust and Genetically Modified Food" conference. London: Institute of Physics. July 6.

Macnaghten, Phil, and John Urry. 1998. *Contested Natures*. London: Sage.

McHughen, Alan. 2000. *A Consumer's Guide to GM Food*. Oxford: Oxford University Press.

Rooker, Jeff. 1997. Interview on BBC-TV *First Sight*. November 27.

Royal Society. 1999. Toxicity of GM Potatoes: Review of Data. Royal Society. June.

Times (London). 1999a. Cap'n Birdseye Puts Freeze on GM Foods. April 28.

———. 1999b. "GM Food Will Return," Says Peer. June 10.

United Kingdom. Cabinet Office. 1999. New Measures on Biotechnology Announced. Press release. May 21.

United Kingdom. Department of the Environment, Transport and the Regions. 1999. Voluntary Agreement on GM Crops Extended. News release. November 5.

United Kingdom. House of Lords. Select Committee on the European Communities. 1998. EC Regulation of Genetic Modification in Agriculture. Evidence given before the select committee. December 15. London: HMSO.

United Kingdom. Ministry of Agriculture, Fisheries, and Food. 1998a. Consumer Choice Wins on Genetically Modified Foods. News release. May 21.

———. 1998b. Government Announces Fuller Evaluation of Growing Genetically Modified Crops. News release. October 21.

Wiener, Jonathan, and Michael Rogers. 2002. Comparing Precaution in the United States and Europe. *Journal of Risk Research* 5:317–49.

PART II

Consequences

Introduction

Antibiotech protestors now jet from one world trade meeting to another dressed as "killer" tomatoes and strawberries with fish heads—fishberries, the protestors call themselves. With signs and vitriolic speeches, they claim to be demonstrating solidarity with the developing world. Capitalist malefactors are accused of trying to foist "Frankenfoods" upon malnourished innocents as part of a nefarious plan to undermine opposition to genetically modified crops and foods.

These campaigns by nongovernmental organizations and protest groups against "Corporate America" and "multinationals" would be entertaining if there were not so much at stake for 800 million desperately hungry people in Africa, Asia, and South America on whose behalf these protestors officiously speak. With the population of the less-developed world set to explode by another two billion over the next thirty years, the surging numbers of the malnourished constitute a tragedy that demands every weapon in the arsenal of the technologically-advanced world, including biotechnology. While the first generation of genetically modified (GM) soybean and cotton crops were ridiculed because they improved production in industrialized nations rather than focusing directly on the needs of consumers or poorer farmers, newer products are attacked even though they bring vast consumer and environmental benefits, offer cost-cutting advantages to even the smallest family farmer, and deliver demonstrable nutritional benefits to millions of the world's poor. There is distressingly little rational discussion about the relative risks of this new technology in comparison with conventional and organic production techniques, which by many measures are far more environmentally problematic.

In this next section, three distinguished commentators examine the causes and real-world impact of the collapse of public discourse on agricultural

78

biotechnology. Andrew S. Natsios, the administrator for the U.S. Agency for International Development (USAID); Robert L. Paarlberg, professor of political science at Wellesley College, Massachusetts; and Carol Tucker Foreman, the director of the Consumer Federation of America's Food Policy Institute, each discuss the infection of food policy by international politics and the cloud that now hovers over the future of agricultural biotechnology.

In "Hunger, Famine, and the Promise of Biotechnology," Andrew Natsios shares his very personal encounter with the horror of famine, which inspired a long and distinguished career in the field of international development. Prior to his present tenure at USAID, Natsios served there first as director of the Office of Foreign Disaster Assistance from 1989 to 1991, and then as assistant administrator for the Bureau for Food and Humanitarian Assistance (now the Bureau of Democracy, Conflict and Humanitarian Assistance) from 1991 to 1993. He poses a question that should ring in the ears of those who opposed the development of Golden Rice or the giving of biotech grain to the developing world: Why has there been a drop in agricultural productivity in Africa at a time when there is a dramatic increase everywhere else in the developing world?

Robert Paarlberg, in "Let Them Eat Precaution," directly addresses the conundrum that Natsios raises. In addition to his teaching responsibilities at Wellesley, Paarlberg is an associate at the Harvard University Center for International Affairs and a consultant with the International Food Policy Research Institute, USAID, the U.S. Department of Agriculture, and the U.S. State Department. He has authored dozens of articles and books, including *The Politics of Precaution: Genetically Modified Crops in Developing Countries* (2001). Paarlberg examines the role of nongovernmental organizations in shaping the public debate and deconstructs the popular myth that American cultural and corporate arrogance is solely to blame for the current gulf between the EU and the United States over biotech policy. "The biggest losers from a continued spread of European-style GM crop regulations," he concludes, "will be poor farmers in the developing world, who will miss the opportunity to enjoy the future productivity gains that might otherwise come from this technology were it to be developed in a less stifling regulatory environment."

Carol Tucker Foreman's "Can Public Support for the Use of Biotechnology in Food Be Salvaged?" suggests that the misgivings over biotechnology run far deeper than its proponents acknowledge. Foreman speaks as an advocate for the consumer, a position she has represented for two decades. Before joining the Consumer Federation of America, she oversaw development of the federal government's first dietary guidelines, served as assistant secretary of agriculture in the Carter administration, and was a member of the U.S. Department of Agriculture's Advisory Committee on Agricultural Biotechnology. She is currently a member of the EU/U.S. Consultative Forum on Biotechnology and the Agricultural Policy Advisory Committee for Trade. Foreman challenges the optimism of GM proponents who believe that new products will dramatically transform public opinion, and instead argues that only stricter regulations will placate jittery consumers. She believes that any tinkering with the food chain is apt to stir anxiety, suggesting a long and arduous road ahead for the agricultural GM industry.

4

Hunger, Famine, and the Promise of Biotechnology

Andrew S. Natsios

I have seen famine up close. A member of my family, my great-uncle, starved to death during the Nazi occupation of Greece in 1942–43, when a half-million Greeks died from famine. Poor to begin with, the country was stripped by the German general staff to feed Rommel's army in North Africa. On top of that was a British blockade. Food did not reach the people, and literally there were bodies in the streets. My great-uncle was buried in a mass grave. The famine scarred the memory of the Greek people and of my own family.

Hunger and famine, to me, are not academic questions. I have taken part in the response to major food emergencies in the world in the last fourteen years. I have seen mass graves.

Today we are facing a serious problem, in Africa in particular, but also to a lesser degree in Central Asia. A third of Africans—almost 200 million people—are chronically food insecure. More than half of Africa's population, about 300 million people, live on less than a dollar a day.

Why has there been a drop in agricultural productivity in Africa at a time when there is a dramatic increase everywhere else in the developing world? Back in the 1980s, the U.S. Agency for International Development (USAID) spent $1.3 billion a year on agricultural development. After administrator Peter McPherson left in 1987 to become deputy secretary of the Treasury Department, that amount was drastically reduced. When I rejoined the agency in 2001, we were spending $243 million. We had dropped a billion dollars, and that's not inflation adjusted.

Africa's reduced agricultural productivity was especially striking because it is increasing dramatically everywhere else in the developing world. But it is the poorest countries that suffer most when aid is cut back. The United States was not alone in cutting back on funding for agriculture. The World Bank, all of the regional banks, the Europeans, and the Canadians followed. The dramatic decline in agricultural aid spending led to decreasing productivity in the poorest areas that are the most food insecure, the most prone to famine.

How bad is the food situation in Africa?

- Basic yields per hectare in Africa are one-fifth of what they are in China.

- Fertilizer use in Africa is eight kilograms per hectare. In Latin America, it is over sixty kilograms, and in Asia, it's one hundred kilograms.

- Only 4 percent of Africa's farmland is irrigated, as compared to 29 percent in the Middle East and 34 percent in Asia. Most of the major famines in Africa in the last thirty years have been driven by drought exacerbated by war.

- The Green Revolution took root in Latin America and Asia, where 60–80 percent of the crop area is planted with improved varieties or hybrids. In Africa, the figure is in the 20–30 percent range.

The success of the Green Revolution in Asia provides a telling contrast to the situation in Africa. Much of it has been due to the wonderful research at the CGIAR, the Consultative Group on International Agricultural Research, a network of research institutions supported by more than sixty donors with the Secretariat at the World Bank. USAID has been the largest donor to the CGIAR network since it was founded in the mid-1960s. Along with the Rockefeller and Ford Foundations, and other bilateral donors, USAID played an important role in helping establish the CGIAR and in supporting the application of the Green Revolution in Asia. With the exception of North Korea and one incident in 1974 in Bangladesh, the revolution prevented famine in much of the continent.

In fact, the economic growth of the economies in Asia started when they began making substantial investments in agriculture. People think growth started with industrialization, but almost all of the countries that took off economically first developed large agricultural surpluses. At the same time, agricultural productivity declined in Africa as a result of the lack of investment, bad policy in many countries, and war. And the trend continues; a study by the International Food Policy Research Institute (IFPRI), the think tank of the CGIAR network, shows that by 2020–26, 16 million more African children will be malnourished if we do not turn the situation around.

But if we can increase annual crop productivity from 1.5 percent to 2.5 percent, the opposite will happen: Malnutrition will decline instead of increase. That is not unfeasible; the productivity increases in Asia were much more substantial than that. And research recently published in *Science* magazine (Evenson and Gollin 2003) shows that the Green Revolution is finally reaching Africa, where improved seed varieties are arriving in West Africa, Uganda, parts of Kenya, Mozambique, and Angola.

The Debate over Biotechnology

Where does biotechnology fit into this picture? As we have all witnessed, it has come under attack, particularly in Europe. It is troubling and even tragic to see the politics and economic competition of the West being imposed on the developing world. Political attacks on biotechnology reached a peak in 2002, just before the Johannesburg Summit on Sustainable Development, causing several famine-ridden countries in southern Africa to initially reject offers of U.S. food aid over concerns about biotech-derived corn.

During this attack over food aid in 2002, the United States was accused of dumping unwanted grain containing biotech, in essence forcing people to accept food that some claimed we would never consume. All of the food aid that we buy in the United States for use abroad is bought on the commercial market. There is an illusion that there are silos that say "Food Aid" or that say "Biotech." That is nonsense; we are not dumping surpluses or dumping biotechnology. When we want food aid, we go to the same commercial market that everybody else goes to and buy food from the food-processing companies. Since biotechnology is fully integrated into the U.S. food system, that food aid may contain biotech. Americans have been eating food derived from biotech crops for nine years. It is integrated fully into our food system as well as those of Canada, Argentina, South Africa, China, and a growing number of countries worldwide. Counter to perceptions, Europe also uses biotech-derived foods and animal feed. Europe imports biotech soybeans for animal feed and vegetable oil from the United States, as well as Argentina and Brazil. The point here is that biotech foods and feeds have been found to be safe by more than a dozen countries around the world and have a track record of safe consumption by millions of people for almost ten years. When we offer food aid, it is the same food we eat at home.

In addition to attacking the motivations of the United States in food aid for Africa, the debate around biotechnology denies any role for this technology in improving productivity of African agriculture. I would like to refute these arguments based on our own experience working in partnership with a dozen African countries and African organizations.

Biotechnology Is Not What Africa Needs; Investment Is. Opponents of biotechnology claim that Africa needs investment in areas other than biotechnology, and we are taking away investment by focusing on it. Investment in areas other than biotechnology is indeed important, but investment in agricultural research, including biotechnology, is critical to the economic development and food security of Africa. As noted, the economic gains in Asia began with gains in agriculture. In Africa, where up to 75 percent of the poor depend on agriculture for income, and where agriculture comprises from 20 to 50 percent of the GDP in many countries, you cannot get economic growth without agricultural growth. Africa is the only region where per-capita spending on agricultural research declined in the last two decades.

At the heart of those agricultural gains is technology embedded in the seed. Since 1980, 50 percent of the increase in agricultural productivity in the developing world has been the result of genetic improvements delivered through seeds. If we're going to increase productivity in Africa, we're going to have to use this technology. It will not happen just by improving traditional methods, but by introducing improved varieties and hybrids. Each country is different, and so are the needs of its agricultural system. One of the answers to the problem of productivity is clearly seed technology, and biotech is a valuable new tool to add to our arsenal.

In South Africa, we can already see the potential. Biotechnology is taking off, not just in Bt cotton, but also in maize production. And it's not just the yellow maize we eat in the United States, but also white maize, of which Americans eat very little, although it is the staple in much of Africa. Following a dramatic rise in the portion of the corn crop in South Africa that uses biotech, as well as the cotton crop, these biotech crops are now an integral part of the South African economy. And it is not just the big farmers who benefit. In the villages, formerly poor black farmers are using this seed, and they're making a lot of money; many have gone from earning a couple of thousand dollars a year to earning ten thousand dollars. Studies present empirical evidence showing that in a couple of areas, there has been a dramatic rise in family income as a result of this technology.

This is not to say other investments are unimportant. USAID invests in a broad strategy for African development, including broad investments in agriculture. But, given the experience of countries like South Africa, China, and India, which demonstrate the economic benefits to small farmers of applying biotechnology, we cannot exclude this among our tools for development.

Biotechnology Will Make African Farmers Dependent on Multinational Companies. USAID's answer to this critique is to say that biotechnology is not only the domain of multinational companies; public research institutions, in Africa and internationally, are actively conducting biotechnology research. The core of USAID's biotechnology strategy is to develop biotech capacity in Africa. The United States has developed a program called the Collaborative Agricultural Biotechnology Initiative (CABIO), which works with American research institutes to link our scientists with African scientists, as well as supporting African research institutions and organizations directly. Through such collaborative research and training, African scientists are already engaging in development of biotechnology applications for their own purposes, in their own countries, according to their own traditions.

In January of last year, we opened a biotech research center in the Ministry of Agriculture in Egypt, which will begin to revolutionize Egyptian agriculture. One South African scientist is developing a corn variety for resistance to drought by taking a gene from a plant that requires almost no water to grow. The South Africans will crossbreed that with a variety of white maize to produce something suitable for the agroclimatic growing situation in southern Africa, which will provide an added buffer to the repetitive tragedy of poor people facing starvation every time there is a drought.

The United States, in collaboration with the World Bank and the Gates Foundation, is also developing biotech foods specifically to address nutritional problems in Africa and other developing countries—an approach known as biofortification. For example, we know that two treatments of vitamin A per child will reduce the death rate from malnutrition by 25 percent, as well as protect against infection, but the cost of a system to get every child taking vitamin A pills twice a year is too high. So instead, we add vitamin A to the food supply, along with zinc and iron, which also drive down the death rates for pregnant women. In collaboration with Monsanto, Iowa State University, the University of Illinois, and two CGIAR centers, we are working on a vitamin A–enhanced corn called Golden Maize for African use. Golden Rice, a high–vitamin A rice, will also be used in Asia. In India, we are developing a public-private alliance to supply vitamin A–enhanced mustard oil, since mustard oil is the principal oil used by poor residents who suffer from high rates of vitamin A deficiencies.

These are all examples of public research for and with Africans. The private sector can offer some valuable technologies to Africa, as the cotton and maize examples in South Africa I mentioned earlier illustrate. Extending that impact to other important staple crops, and to farmers not served by multinational seed companies, is the role of the public sector. This is not a case of wishful thinking, but a case of opening one's eyes to the fact that African research institutions are already conducting biotechnology research.

Biotech Foods Force African Muslims to Violate Their Religious Dietary Rules. According to a government minister in one southern African country, opponents of genetically modified organisms (GMOs) are telling people in Muslim villages that the Americans have put pig genes in the American corn that we donate for food aid—a violation of Islam for those who eat the corn.

"Is that true, Andrew?" the minister asked me.

No, it is not. I know all of the corn varieties that are biotech or approved varieties, and none of them has animal genes of any kind in it. Ethical and cultural acceptability is important in the United States as well as overseas, as we have a diverse population, including Muslims. This is irresponsible scaremongering

American Biotechnology Will Jeopardize African Export Trade to Europe. I was also approached by the minister of agriculture in another African nation, who said that antibiotech activists had urged him to reject U.S. food aid maize—this in the middle of a major drought, a very serious situation where malnutrition rates had been rising alarmingly. "They're telling us if we accept your food aid we won't be able to export to Europe," the minister told me. Although the African countries do not export their maize to Europe, they do export coffee, tea, nuts, vegetables, and fruit. Warned by activists that their farmers will plant the seeds from the American corn sent to their countries as food aid and that the plants will cross-pollinate, Africans are reluctant to accept donations of yellow corn that they think will crossbreed with their vegetables, fruits, and nuts, making this produce unacceptable in Europe. This is pure hysteria, based on bad science. You cannot have open pollination of corn with fruits and vegetables. You can cross-pollinate one variety of corn with another variety of corn, but you cannot do it across different crops. It is scientifically impossible.

Biotechnology-Derived Crops Will Adversely Affect the Environment in Africa. The concern has been raised that if food aid recipients were to plant U.S. biotech maize, it could result in negative impacts on the environment. First, there are no wild plants, or native biodiversity, with which maize can cross-pollinate in Africa. Second, the South African government undertook environmental reviews of biotech maize, as required by their national law, before approving the planting of several different types of biotech maize. In addition, USAID has funded environmental research on the potential impacts of insect-resistant or bt maize in Africa and found no evidence of undue risk. We fund other such research to look at the potential implications of other biotech crops such as cowpea and sorghum for

Africa as part of our commitment to ensuring the safe and effective use of this technology. As these examples illustrate, African researchers and regulatory officials have and are continuing to review the environmental concerns in the African context.

Returning to the argument surrounding food aid, the yellow maize that the United States sends as food aid is inappropriate for African farmers and thus is unlikely to be used for planting for any period of time. For one thing, it won't grow well in African climatic conditions. For another, Africans are not about to begin growing yellow corn because they do not particularly like it. We offer it to them because it is what we have available and it saves people from dying of starvation in a famine, but it is not likely to become a staple of the African diet in more plentiful times. Moreover, when people are hungry, they don't plant their seeds; they eat them, particularly if they don't think they're going to live to the next harvest. They even eat the seeds that we give them to plant. This is a serious problem—people eat all of their seed stock, and there's no seed stock left. This is a quandary we're facing in Ethiopia right now.

Placing Biotechnology Concerns into Context

Looking back at the spillover from the U.S.-European biotechnology debate over food aid, the rhetoric of those opposed to the technology unfortunately masks the fact that there is substantial common ground between the United States and European development agencies on both the safety and the potential benefits of biotechnology for developing countries. During the crisis over the safety of food aid in southern Africa, the EU officials issued statements to reassure Africans that their own reviews of the safety of a number of varieties of biotech corn had determined them to be safe and that all trade with Europe was not at risk if they were to accept food aid that contains biotech. The EU and some European donors have also considered the potential benefits of applying biotechnology to improve agricultural development and nutrition, clearly leaving open this option for developing countries. Attacks on biotechnology seek to raise alarms out of proportion with risk and to suggest more controversy and division than we see either in Africa or in working with Europeans on African development.

As another example of this, I was particularly disturbed by a handout that circulated during the Johannesburg Summit. It originated with a group in Britain and read, "If you eat biotech food, your DNA will be altered."

I called in one of our chief biotech officers and asked him: "Is that true?"

He said, "It's ridiculous." Yet the flier was sent out en masse from the Johannesburg Summit and was widely believed.

Are there legitimate concerns surrounding agricultural biotechnology? Absolutely. There are risks to all technology. But they need to be put into context. The debate over the use of pesticides offers an informative analogy about the risks and rewards of technology. We have a serious problem with the misuse of pesticides in the developing world. Why are pesticides still used? Because sometimes they are necessary. For example, during the first Bush administration, northern Africa experienced a terrible locust plague that was going to cause a famine. Julia Taft, my predecessor at USAID, now at the United Nations Development Programme (UNDP), launched a massive antilocust campaign using pesticides. It stopped the destruction of crops and stopped the locusts.

However, in less dramatic and less threatening situations, particularly in countries that cannot use them responsibly, it is much better not to use pesticides. The amount of poisoning in the developing world is very serious, with people dying because pesticides are not used properly. In China, farmers even buy atropine, which prevents certain kinds of poisoning from taking place. They give it to their kids when they're using pesticides on the farm.

Here is a case where biotechnology can reduce risks to both the environment and human health. The United States is addressing these concerns by developing insect- and virus-resistant varieties of grain that can avoid the use of pesticides. Biotechnology has clearly demonstrated benefits, not just risks.

The final argument is that we should not decide for Africa, but listen to the diversity of voices and allow Africans to decide. We hear many voices in our collaborations with African institutions that want the opportunity to join the Green Revolution. In a recent report issued by the Forum on Agricultural Research in Africa, they declared "their commitment to . . .

building Africa's human and physical capability in biotechnology to be able to engage with global public and private sector partners to capture the advances needed to sustainably intensify African agriculture" (Forum on Agricultural Research in Africa 2003). Biotechnology is not going to solve all of Africa's problems, but it will solve some of them. It is a tool that can address hunger, malnutrition, and even development. We believe we need to encourage Africans to develop this technology and use it as they see fit.

References

Evenson, R. E., and D. Gollin. 2003. Assessing the Impact of the Green Revolution, 1960 to 2000. *Science* 300, no. 5620:758–62.

Forum on Agricultural Research in Africa. 2003. *The Dakar Declaration.* Dakar, Senegal: FARA. May 20.

5

Let Them Eat Precaution:
Why GM Crops Are Being Overregulated in the Developing World

Robert L. Paarlberg

The surprising power and reach of Europe's influence over regulatory policies concerning genetically modified (GM) agricultural crops in the developing world first came to light in the summer of 2002, when a number of states in southern Africa, all facing possible famine conditions, nonetheless began rejecting GM corn from the United States as food aid. They did so partly in response to European officials who had taught them to follow the "precautionary principle" toward GM technologies, and partly in response to European-based nongovernmental organizations (NGOs) that had told them GM crops were unsafe for human health and the environment.

But they also had a commercial motive; they feared that if some of the GM corn kernels were planted, African exporters might be denied sales in the European Union market, just as had already happened to exporters of GM corn from the United States. The food aid being offered in 2002 was the same GM corn Americans had been consuming since 1996 and the United Nations World Food Programme (WFP) had been distributing in Africa over the previous six years, but now Zambia rejected it completely, and three other countries—Zimbabwe, Mozambique, and Malawi—accepted it only if it were milled first to prevent planting.

These events persuaded trade officials in the United States that the highly precautionary policies toward GM foods that had taken hold in

Europe after 1996 were now spreading to the developing world and becoming a threat, not just to U.S. farm commodity exports, but also to agricultural development in poor countries and to the efficient international movement of food aid for famine relief. In 2002–3, the WFP was able to replace most U.S. corn shipments to Zambia with non-GM corn from Tanzania and South Africa, but not before hardships in the country increased. In January 2003, a mob of 6,000 hungry villagers in one rural town in Zambia overpowered an armed policeman to loot a storehouse filled with U.S. corn, in the knowledge that the government was soon going to insist it be taken out of the country.

U.S. frustration is deepened by an awareness that a great many officials and scientists in Europe share the view that GM foods are safe. In 2001, the Research Directorate General of the EU released a summary of eighty-one separate scientific studies, all financed by the EU rather than private industry, conducted over a fifteen-year period and aimed at determining whether genetically modified products were unsafe, insufficiently tested, or underregulated. None of these studies found any scientific evidence of added harm to humans or the environment from any approved GM crops or foods (Kessler and Economidis 2001).

In December 2002, even the French Academies of Sciences and Medicine drew a similar conclusion. The academies issued a report that said "there [had] not been a health problem . . . or damage to the environment" from GM crops. This report blamed the rejection and overregulation of GM technologies in Europe on what it called a "propagation of erroneous information" (French Academy of Sciences 2002). In May 2003, the Royal Society in London presented to a government-sponsored review in the United Kingdom two submissions that found no credible evidence that GM foods were more harmful than non-GM foods (Royal Society 2003). In March 2004, the British Medical Association (BMA) endorsed this finding (British Medical Association 2004).

The United States may have a sound scientific and legal case against the restrictive policies set in place by the EU since 1998, but the political and commercial foundation for challenging the EU on GM foods and crops is quite weak. In the global fight over GM crop regulation, the political and commercial influence of the EU now exceeds that of the United States. Despite U.S. challenges in the World Trade Organization (WTO), highly precautionary EU-style regulations on GM foods and crops will probably continue to spread to much of the developing world. If so, the biggest losers will not be commercial farmers or GM crop exporters in the United States; they will be poor farmers in developing countries, who will be denied new GM options to overcome serious constraints on farm productivity.

Reasons for the Restricted Planting of GM Crops

GM seeds have been commercially available since 1995, yet 96 percent of all the world's plantings of GM crops are still restricted to just five countries— the United States, Canada, Argentina, Brazil, and China (James 2004). This restricted spread of GM crops in part reflects a globalization of Europe's highly precautionary regulatory approach toward this technology. In most developing countries, it is still not legal for farmers to plant any GM crops. Critics often depict globalization as tantamount to Americanization, since it so often reflects the global spread of American tastes, regulatory preferences, and technologies. So why, in this case, are we seeing European tastes and regulatory preferences triumph? Let us consider the four different channels of influence through which this Europeanization of regulatory standards is now taking place: intergovernmental organizations, development assistance, nongovernmental organizations, and international food and commodity markets.

Intergovernmental Organizations. It is unsurprising that European influence dominates within most of the international intergovernmental organizations (IGOs) that currently deal with GM foods and crops. European governments work hard to maintain and develop their influence within IGOs, while the U.S. government too often ignores or disrespects them by failing to send high-ranking delegations to IGO meetings, or ratify conventions, or pay dues on time. In part as a consequence, some IGOs that should be promoting GM crops are not doing so, while the IGOs that are regulating GM crops are doing so in the precautionary manner that Europeans prefer.

International agricultural organizations and development organizations such as the UN Food and Agriculture Organization (FAO), the Consultative Group on International Agricultural Research (CGIAR), and the World

Bank should be promoting GM crops because these organizations have traditionally been production-oriented and protechnology. But in the current climate of European misgivings toward GM crops, the organizations have all been reluctant to promote them.

The FAO, headquartered in Rome, said nothing positive about genetically modified organisms (GMOs) prior to 2004, and FAO's director general had even stated publicly that GMOs would not be needed to meet the objective of alleviating world hunger by 2015. A more positive tone finally emerged in May 2004, when the FAO's annual publication, "The State of Food and Agriculture," concluded that GM crops might indeed be able to play a role in meeting the needs of the poor (Food and Agriculture Organization 2004). Yet the response to this report was so hostile—670 separate NGOs wrote a letter to the FAO calling the report a "stab in the back" to farmers and the rural poor—that this UN agency has now fallen back into silence on the topic.

Nor is the CGIAR system promoting GM crops. It is true that the International Rice Research Institute (IRRI) in the Philippines is supposed to be developing "Golden Rice." But that will be difficult, since the institute has decided not to conduct any field trials of GM crops in the Philippines, lest they stir up the anger of local anti-GM NGOs. Only two out of IRRI's eight hundred scientists have recently been working on Golden Rice (MacIntosh 2001). IRRI is in a position to go ahead with several other GM rice innovations, including iron-enriched "dream rice," good for iron-deficient women and children, and "aerobic rice," which requires less water to grow; but the scientists working on these innovations face extensive critical scrutiny from anti-GM NGOs and from European donors.

Elsewhere in the Consultative Group (CG) system, Mexico-based International Wheat and Maize Improvement Center (CIMMYT) is participating in an insect-resistant maize project for Africa, but the major funding comes not from CIMMYT, but from the Syngenta Foundation. CIMMYT has also been testing wheat plants genetically engineered to withstand drought, but resistance in Mexico from environmental groups has induced caution. Following the discovery in 2001 that some GM corn had been planted in Mexico, the Mexican government placed tighter constraints on all GM crop research in the country, including at CIMMYT. It is indicative that the 2003 CGIAR annual report makes no reference to GM crops. This reticence is no surprise, given that European contributions to the CG budget are now twice as large as contributions from North America. Those who pay the piper call the tune.

Fear of diminished European financial support and of criticism from European NGOs has also paralyzed the World Bank. In 1999, the World Bank attempted to draft a strategy document on GM crops, but because of political opposition at the top, this paper, as bland as it was, was never even presented to the board for approval inside the Bank. Now the World Bank's strategy on GM crops is not to promote them, but to study them. In late summer 2002, the Bank announced a three-year global consultation process designed to examine the "possible benefits" of this new technology, and the alleged drawbacks. There is no danger of running into any criticism from the EU with this approach. On GM crops, the Bank now spends more time worrying about and responding to NGO criticisms than imagining how to advance this technology for the benefit of poor farmers in the developing world.

While most of the IGOs that might be promoting GM crops are not doing so, a number of equally powerful ones are taking a distinctly European approach toward regulating this new technology. For example, the United Nations Environment Program (UNEP) is now using funds from the Global Environment Facility to help developing countries draft precautionary biosafety regulations for GM crops. UNEP wants such regulations to be in place before these countries begin any planting of GM crops, whatever delay or administrative cost this might imply.

Also within UNEP's Convention on Biological Diversity, governments with European encouragement negotiated in 2000 a new Cartagena Biosafety Protocol, which took effect in September 2003 after being ratified by one hundred countries. This protocol explicitly endorses "the precautionary approach" toward regulation of living GMOs. The protocol imposes informational obligations on exporters of living GM plants and seeds, and allows importing governments to restrict imports even without scientific demonstration of a specific risk to the environment. The protocol states that under conditions of scientific uncertainty no government should be prevented from blocking imports of GM plants or seeds. From a scientific standpoint, of course, there will always be uncertainty. Testing for the nth hypothetical risk and then for the nth year of exposure to that nth risk is not a requirement we impose on any other product, because it so easily becomes a formula for endless delay.

The terms of the Cartagena Protocol were modeled after an earlier pact, the Basel Convention on Transboundary Movement of Hazardous Wastes. Treating GM crops like a hazardous waste suggests that a bias against the technology was part of this new agreement from the start. American influence over the negotiations was undercut because the U.S. Senate had never ratified the original Convention on Biological Diversity, and since the United States was not party to the convention, it had to participate in the protocol negotiations as an "observer"—not a good way to shape the outcome.

Development Assistance. Development assistance is a second channel through which European influence over GM crop regulations is now being extended internationally. During the Cold War, the United States financed assistance generously, hoping to win friends and allies in the developing world. With the end of the Cold War, U.S. assistance programs have now withered, particularly in the area of agriculture. Between 1992 and 1999, support from the U.S. Agency for International Development (USAID) for agricultural development assistance fell by more than 50 percent. In Africa, the United States largely withdrew from agricultural development assistance work. Agricultural specialists were no longer sent to the field, USAID missions were closed down, and people were brought home. Meanwhile, European donors remained very much on the scene, ready to advise African governments on how to regulate GM crops. The Dutch, Danes, and Germans remained active, consistently advocating ratification of the new Cartagena Protocol and formal adoption of a Europe-style precautionary approach.

Developing countries are warned not to plant GM crops until highly precautionary biosafety screening procedures are in place. The implications of this advice can be seen in Zambia, where the absence of adequate biosafety regulations was one reason given by government officials for rejecting GM food aid from the United States in 2002–3. This regulatory deficit will soon be corrected, at a dubious price. In April 2003, Zambia's Ministry of Science and Technology announced a five-year, $40 million National Biosafety and Biotechnology Strategy Plan designed not so much to develop useful GM crop technologies for Zambia's struggling farmers as to regulate those technologies so as to protect the biodiversity of the country (Hanyona 2003). The government of Zambia knows that European donors will pay for this expensive plan. Instead of exporting useful GM technologies, the European donor community is exporting GM regulations.

The practical result in many countries has been regulatory paralysis. Once strict biosafety screening requirements have been written into the laws, cautious politicians and bureaucrats discover that the safest thing, politically, is to give no GM crop approvals at all. Approving nothing is one way to conceal a weak technical capacity to screen GM technologies on a case-by-case basis, and the best way to avoid criticism from NGOs or difficult questions from the media.

This is one reason so few biosafety approvals for GM crops have been given by developing countries. Other than South Africa, not a single country on the African continent has yet approved any GM crops for commercial planting. Other than the Philippines, not a single country in all of developing Asia has given even one biosafety approval for the planting of any major GM food or feed crop—no corn, no soybeans, no rice. Insect-resistant GM cotton has been approved in a few Asian countries, including China, India, and Indonesia, but no GM food or feed crops. We sometimes hear the question, "If this technology is so good, why aren't more poor farmers in the developing world planting GM crops?" The reason is their own government regulators have not yet made it legal for them to do so (Paarlberg 2001).

European NGOs. European-based NGOs are another source of external influence over GM crop regulations in poor countries. Environmental and antiglobalization NGOs have invested heavily in an effort to block this new technology. These NGOs were instrumental in forcing the EU to impose a moratorium on new GM crop approvals after 1998, and now they are working to prevent approvals in the developing world. Greenpeace, which has an annual budget of more than $100 million, has invested heavily in a global campaign to stop genetic engineering, focusing particularly on developing countries that have not yet approved any GM crops.[1]

Of course, private biotech companies like Monsanto spend a lot more money to promote the spread of GM crops. But NGOs can go beyond paid media campaigns; they also employ direct actions, street protests, and lawsuits to generate free media attention. NGO lawsuits have emerged as a proven method for delay in poor countries. In 1998, Monsanto thought it had won official approval in Brazil for five varieties for Roundup Ready soybeans. A local consumer NGO and the Brazilian office of Greenpeace filed a lawsuit and found a sympathetic federal court judge to issue an injunction that stopped the approval. This case became caught up in the Brazilian court system, and as late as 2003 it remained technically illegal to plant any GM seeds in Brazil, even though farmers there were doing so, having smuggled in seeds from Argentina.

In India, when a local partner of Monsanto began conducting field trials to attain biosafety approval for Bt cotton, NGOs with European links invaded the trials, uprooted the cotton plants, and burned them. The Indian government had been told by the NGOs that the plants contained a so-called terminator gene that would render the seeds of plants sterile. This was not true, but headlines were made, public interest litigations were filed, and the approval of GM cotton was delayed for two years. In 2002, India went ahead with GM cotton in the southern part of the country only, but it held back from approving GM mustard or any other food or feed crops.

An NGO grip on Indian GM crop approvals reappeared in April 2003, when the government's Genetic Engineering Approval Committee (GEAC) rejected an Indian seed company's request to extend the sale and planting of GM cotton seeds into the northern part of the country. GEAC denied this request despite the success of GM cotton in the south and despite a lack of evidence of any particular biosafety risk linked to GM cotton in the north.[2] Environmental groups in India led by Greenpeace India and the Research Foundation for Science, Technology and Ecology (RFSTE) hailed this as a major victory for their anti-GM campaign.

European-based NGOs are also working hard to keep GM varieties out of food aid shipments to the developing world. Oxfam International, based in the United Kingdom, has joined Greenpeace in urging the governments and the FAO to develop and implement standards that prevent the distribution of GMOs in food aid (Sharma 2003). NGO campaigns against GM foods and crops played a visible role in the slowing of GM food aid deliveries to southern Africa in 2002–3, at a time when roughly 15 million people in six drought-stricken countries faced a severe hunger crisis. Political leaders in this region had been frightened away from GM in part by NGO campaigns conducted by groups such as Action Aid from the United Kingdom and Friends of the Earth from the Netherlands. In 2002, the government of Zambia refused even to take milled corn from the United States, citing food safety concerns and invoking the precautionary principle. This import ban was reaffirmed in November 2002, after a team of Zambian experts traveled to Europe and North America to seek advice on the issue. Among the experts consulted in London, Brussels, and Amsterdam were NGO leaders from Greenpeace, Friends of the Earth, and other organizations deeply opposed to GM technologies. They were also influenced by the views of the British Medical Association, which had no evidence of any added health risks from GM foods but was still clinging at that time to a position it had taken in 1999, that the technology had not yet been sufficiently tested for all hypothetical risks (*Observer* 2003).

Frustrated with this crisis, United States officials at one point late in 2002 asked the EU and the World Health Organization (WHO) to reassure officials in southern Africa that there was no scientific evidence of risk from the corn being offered, and to remind the Zambians that even EU regulators had given food safety approvals to some varieties of GM corn and soybeans. The first response from the EU was to say this was a matter between the United States and Zambia. WHO Director General Gro Brundtland had earlier disappointed the United States by saying to a group of health ministers from southern Africa that the GM corn was "not likely" to present a risk.[3]

International Markets. International markets are a fourth and final channel through which European attitudes toward GM foods and feeds are now spreading beyond Europe. It was originally assumed that once the United States began growing GM food and feed products, the technology would quickly become pervasive, and hence impossible to resist in international markets. The United States is the world's biggest exporter of agricultural goods, so it seemed these products would have to be accepted worldwide. That was the wrong way to look at the matter. In international commodity markets, it is the big importers, not the big exporters, who usually set standards. In any competitive market, it is the customer who is always right.

In commodity markets, the key customers are the biggest importers, led by Europe and Japan. The EU and Japan together take in $90 billion worth of agricultural imports every year. From developing countries specifically, the EU is the big importer. Europe alone imports 75 percent more farm products from developing countries every year than the United States. The

EU imports more food and farm products from developing countries than the United States, Japan, Canada, and Australia combined. Accordingly, developing countries that aspire to export farm products must pay close attention to European consumer preferences and import regulations.

If it were only consumer opinion in Europe now weighted against GM foods, nobody could really complain, since that would be a free-market outcome. Increasingly, however, consumer opinion alone is not the obstacle to GM product imports. Imports are now also being blocked by a variety of official EU regulations and policy actions, some of which derive from political rather than pure consumer preference. European consumers have shown they are willing to pay premiums in the market for GM-free food, but only small ones. Politicians and regulatory bureaucrats in Europe have nonetheless set in place rules that could effectively remove GM products from the shelf completely. European food safety regulators, having underregulated bovine spongiform encephalopathy (BSE), dioxin, and hoof-and-mouth disease in the recent past, have been hoping to restore their credibility in part by overregulating GM foods. Without any scientific evidence of risk, the EU in 2004 went ahead with a system for the tracing and labeling of GM products that might drive them out of the EU market completely.

These new regulations go beyond the 1998 informal EU moratorium on new GM biosafety approvals, a policy that effectively blocked roughly $300 million in U.S. corn sales to Europe every year because of new GM varieties in those bulk shipments that could not be approved in Europe. This is a relatively small export loss compared to the volume of sales placed at risk by the traceability and labeling regulations that began taking effect in April 2004. These regulations impose costly new product-segregation requirements on all exporters of GM food or feed products to Europe. Mandatory GM labeling is now extended to animal feed as well as to human food, and also to processed products even where there is no longer any physically detectable GM content.

The EU claims these regulations have been necessary to ease the concerns of consumers, and to facilitate a final lifting of the 1998 moratorium. For exporters, these new regulations could be worse than the moratorium. Fraudulent claims are likely to be made for the GM-free content of processed products, and if strictly enforced, the traceability regulation could drive labeled GM products out of the market completely in Europe. This regulation will oblige every operator in the food chain to maintain a legal audit trail for all GM products, recording where they came from all the way back to the farmer who planted the seed. Purportedly, this will facilitate enforcement of the new labeling rule and make possible a quick removal from the food chain of any GM product that might prove unsafe (Commission of the European Communities 2001). In practice it will discourage many exporters to Europe from planting GM in the first place.

This new traceability requirement does not require much added physical segregation of products on farms in Europe, because farmers there are not growing any GM crops to speak of. For Europeans it will create legal tangles and burdensome paperwork, but perhaps nothing more. Not so in the United States, or in other countries where GM crops are grown and exported to Europe. Depending on how strictly Europe enforces the regulations, such exporters may now have to present documentary audit trails for any GM product they wish to sell in Europe, which means they have to start segregating GM from non-GM products physically at home, and tracing GM products through the market with legal records as well. Segregating GM from non-GM bulk commodities at a low threshold of permitted contamination can require having two of everything—two sets of grain elevators in every county, two categories of railroad cars and river barges, separate drying and processing facilities, and segregated export elevators. At the very low threshold of contamination to be permitted under the new EU regulation (shipments will now have to be 99.1 percent free of any GM content to escape the label and traceability requirement), this segregation process will be costly enough to raise the export price of commodity shipments from GM countries, resulting in a loss of price competitiveness and probably also a loss of importer and consumer acceptance in Europe once GM products are labeled as such. Developing countries lacking the legal and infrastructural capacity to segregate GM from non-GM products will now face an even greater incentive to remain GM free.

Trying to Make a Case against the Regulations

Both the 1998 EU moratorium on new GM approvals and the new labeling and traceability regulations appear to violate key agreements of the WTO. The moratorium has clearly violated the Sanitary and Phytosanitary (SPS) agreement because it has not been based on any scientific evidence of risk. When the United States finally initiated a formal challenge to the moratorium in 2003, through the WTO's dispute-settlement process, the EU Commission responded in 2004 by finally approving, for import but not for planting, two new varieties of GM maize, allowing the EU to claim that the moratorium was no longer in effect. Yet these two new approvals had to be made over the heads of a blocking coalition of EU member governments, a practice not certain to be continued under the tenure of the new commission that began in October 2004.

The new traceability and labeling regulations probably violate the Technical Barriers to Trade (TBT) agreement of the WTO under both articles 2.1 ("like products") and 2.2 ("legitimate objective"). The U.S. government worked hard in 2001–2 to influence the drafting of these EU regulations, hoping to weaken their likely impact on American exports. The United States asked that the threshold of permitted contamination be raised, that labels be required only if some GM content is physically detectable, and that sellers be allowed to say when labeling exports that products or shipments "may contain" certain GMOs, rather than having to say exactly which GMOs they are. But these requests found little support when the EU councils of agricultural and environmental ministers approved the new regulations late in 2002. Only the United Kingdom appeared ready to call for any weakening of the regulations, while the French, at the other extreme, wanted GM labels not only on all processed GM foods, but also on meat from animals that had been fed GM crops, and even on GM pet foods.

Whereas the EU's unofficial moratorium on new approvals may be subject to an effective WTO challenge, the 2004 rules on labeling and tracing are likely to prove far more durable. These rules were legislated as joint regulations of the EU Council and Parliament, under the codecision procedure of the 1993 EU treaty. They come into law at the national level automatically and will be more difficult to reverse than the informal moratorium on new approvals. Even if the United States should eventually win a dispute-settlement challenge against these new rules in the WTO, the result may still be no change in European policy. A three-person WTO dispute panel will carry little political weight with the European Council and Parliament, since the WTO is not a food safety organization, does not embrace Europe's precautionary principle, and can be attacked as nondemocratic.

The United States learned the limits of WTO dispute-settlement powers in the food safety area with the outcome of an earlier challenge to an EU ban on sales of beef from animals raised with growth hormones. This ban was undertaken without any scientific evidence of risk, so the United States challenged its legality under the SPS agreement and won a victory in the WTO's dispute-settlement body in 1998, but still failed to open the EU market to hormone-treated beef. The hormone ban was popular with consumers in Europe, so the council decided to keep it in place and comply with WTO rules by allowing the United States to impose retaliatory tariffs on $117 million worth of EU food exports, hardly a victory for open trade (Roberts 1998). If the United States were eventually to win a case in the WTO against the traceability and labeling regulations, the EU could again decide to accept retaliation rather than comply. Moreover, the spectacle and publicity surrounding a WTO challenge might drive European consumers, and private importers, even farther away from GM products, and perhaps other U.S. products as well.

The regulatory movement in Europe toward tighter restrictions on GM imports is almost certain to continue, and it is now the single greatest inducement for so many governments around the world to keep their farm sectors GM-free. The minute they start planting GM food or feed crops, they might have to set in place the costly product-segregation systems required to comply with Europe's traceability and labeling regulations, or else lose access to Europe. This was one reason for Africa's rejections of GM food aid. Several years ago, private importers in Europe said no to purchasing beef from Namibia because it was partly raised on GM corn grown in South Africa. When Zambia officially reaffirmed its ban on GM food aid in November 2002, it included among its reasons a fear of lost export sales to Europe.

This same fear is now slowing down the technology even in some strongly pro-GM countries, such as Argentina. Following the 1998 EU moratorium, Argentina made it a policy not to approve any new GM varieties until they had been approved for import into the EU. China, another early GM enthusiast, has similarly decided to hold back on the commercial approval of the planting of GM maize, soybeans, or rice. China began its slowdown on GM crops after a shipment of soy sauce made in Shanghai from U.S. soybeans was turned away at the EU border because of its possible GM origin.

Even the United States and Canada are slowing down some new GM crop technologies, such as GM wheat, for fear of losing export sales of wheat or flour in Europe or Japan. In May 2003, South Korean wheat millers warned U.S. wheat producers and government officials that consumers in Korea would boycott all American wheat if GM varieties were introduced. Korean brewers and corn syrup manufacturers had earlier turned away from purchasing U.S. corn, which is known to be GM, switching to Brazilian and Chinese supplies instead. In Canada, the Wheat Board estimated that 82 percent of its foreign customers do not want to buy GM wheat, and asked regulators to take these market realities into account when deciding whether to approve the wheat for commercial use. In 2004, the Monsanto Company announced that because of these commercial acceptance issues, it was not going to seek regulatory approval in the United States or Canada for GM wheat. U.S. acreage planted to GM maize and soy continued to increase through 2004, but the new EU regulations on tracing and labeling could put the future growth in GM crop acreage in all exporting countries under question.

Who Will Lose If the EU Wins?

The damaging trends described here do not have to be a calamity for most farmers in the United States. Even in the extreme case, if U.S. producers were to have to retreat from planting food and feed crops such as GM corn or soy, cotton farmers would not be prevented from continuing to plant GM seeds, and producers of other crops would not be required to make any changes at all. U.S. farm income would dip by several percentage points because corn and soybean production costs would increase, and the spraying of insecticides and herbicides would also increase, perhaps with adverse environmental and occupational safety effects. Yet by one estimate, eliminating GM crops in the United States might imply no more than a 3 percent reduction in annual net farm income (National Center for Food and Agricultural Policy 2004). If the United States were to back away from planting GM corn or soybeans, the biggest commercial losers would be the companies that originally developed these crops, not the farmers currently growing them. Some U.S. food companies will also be affected by the new EU regulations, but they will be able to respond either by reformulating their products to eliminate maize and soy ingredients, or by contracting

Notes

1. Greenpeace is active even in China, having extended its presence from a headquarters in Hong Kong to an office in Beijing. The organization has close contacts with the Nanjing Institute of Environmental Sciences, which even cosponsors its newsletter. This institute falls under the authority of the State Environmental Protection Administration, which would like to be able to exercise greater control over GM crops inside China, where the agricultural and science ministries have traditionally had the lead role.

2. In 2002, Indian cotton farmers in Andhra Pradesh bought enough seed to plant 8,300 acres with GM cotton. By 2003, farmers had booked enough seed purchases to plant 60,000 acres.

3. United States Mission to the EU, August 28, 2002, http://www.useu.be/Categories/Biotech/Aug2802WHOFoodDonationsAfrica.html.

References

British Medical Association (BMA). 2004. Genetically Modified Foods and Health: A Second Interim Statement. London: British Medical Association. March.

Commission of the European Communities. 2001. Proposal for a Regulation of the European Parliament and of the Council Concerning Traceability and Labeling of Genetically Modified Organisms and Traceability of Food and Feed Products Produced from Genetically Modified Organisms and Amending Directive. 2001/18/EC. Brussels: Commission of the European Communities. July 25.

Consultative Group on International Agricultural Research (CGIAR). 2001. Global Knowledge for Local Impact: Agricultural Science and Technology in Sustainable Development. CGIAR annual report. Washington, D.C.: The World Bank CGIAR Secretariat.

Food and Agriculture Organization (FAO) of the United Nations. 2004 The State of Food and Agriculture 2003–04: Agricultural Biotechnology: Meeting the Needs of the Poor? Rome: FAO.

French Academy of Sciences. 2002. GM Plants: Reporting on the Science and Technology. http://www.academie-sciences.fr.

Hanyona, Singy. 2003. Zambia Develops Biosafety Strategy. *Environmental News Service*. April.

James, Clive. 2004. Preview: Global Status of Commercialized Biotech/GM Crops: 2004. ISAAA Brief 32-2004. Ithaca, N.Y.: International Service for the Acquisition of Agri-Biotech Applications.

Kessler, Charles, and Ioannis Economidis, eds. 2001. *EC-Sponsored Research on Safety of Genetically Modified Organisms: A Review of Results*. Luxembourg: Office for Official Publications of the European Communities.

Macintosh, Duncan. 2001. Letter to the Editor. *Business Week*. February 26.

National Center for Food and Agricultural Policy. 2004. Impacts on U.S. Agriculture of Biotechnology-Derived Crops Planted in 2003—An Update of Eleven Case Studies. October. http://www.ncfap.org/whatwedo/biotech-us.php.

Natsios, Andrew S. 2003. Testimony prepared for Senate Appropriations Subcommittee on Foreign Operations. June 5.

Observer. 2003. No Risk in GM Food, Say Doctors. May 25.

Paarlberg, Robert L. 2001. *The Politics of Precaution: Genetically Modified Crops in Developing Countries*. Baltimore: Johns Hopkins University Press.

Roberts, Donna. 1998. Preliminary Assessment of the Effects of the WTO Agreement on Sanitary and Phytosanitary Trade Regulations. *Journal of International Economic Law* 1 (September): 377–405.

Royal Society. 2003. Royal Society Submission to the Government's GM Science Review. Policy Document 14/03. http://www.royalsoc.ac.uk/displaypagedoc.asp?id=11476.

Sharma, Ashok. 2003. NGOs Oppose GM Food Dumping in Iraq, Favour UN Role. *Financial Express* (India). April 14.

U.S. Department of State. 2002. USAID's Natsios on Plan to End Hunger in Africa. Fact sheet. August 30.

U.S. Trade Representative. 2003. U.S. and Cooperating Countries File WTO Case against the EU Moratorium on Biotech Foods and Crops. Press release. May 13.

6

Can Public Support for the Use of Biotechnology in Food Be Salvaged?

Carol Tucker Foreman

Unlike Europeans, most Americans clearly are not profoundly opposed to agricultural biotechnology. But a decade after food biotechnology was introduced in the United States, it is equally clear that Americans have not embraced it. Most are not aware that they've been eating foods produced through agricultural biotechnology for several years, nor are they persuaded that the products are safe. They want government to employ a rigorous regulatory system that protects public health, and they want genetically modified (GM) food products labeled so they can choose to avoid them. The percentage of people who expect to benefit from agricultural biotechnology has declined over the past several years. And while the public opinion jury is still out on acceptance of genetically modified plants, Americans are opposed to having meat and milk from cloned and transgenic animals in the food supply.

Up to now, farmers, the biotechnology industry, food processors, and government regulators have discounted consumer concerns about food biotechnology. It may be unwise for them to continue to do so. The next generation of genetically modified products will be much more visible to consumers than Bt corn and Roundup Ready soybeans. Some of the new products, such as cloned and transgenic animals, evoke a visceral negative response.

Many of the public's concerns could be addressed without causing significant loss to the industry. The question is, will government and industry act before some event, such as the accidental appearance of drug-laced cornflakes or bacon from transgenic or cloned pigs on the family breakfast table, changes concern to outrage?

Food and Feelings

The agricultural biotechnology industry and government regulators argue that we should stop worrying and learn to love genetically modified food because the current products are based on "sound" science, are good for farmers, and are safe for consumers and the environment. They ignore the powerful cultural and personal attachment that most people have to their food. From the apple in the Garden of Eden to the golden arches, food—what, how, and how much we eat—has played an elemental role in human history. We eat to live, but we also live to eat. Food is more than fuel for the body. Because what we eat literally becomes part of our bodies, food is the source of some of our greatest pleasures and, not surprisingly, our greatest fears (Rozin 2000).

Food plays an extraordinarily powerful role in how we live (Visser 1986). Throughout history, people and cultures have been distinguished by what they are obligated or forbidden to eat. Jewish and Islamic dietary laws prohibit eating pork. Hindus abstain from beef. Christians forgo certain foods during Lent. Beans were a forbidden food in the ancient Pythagorean cult to which Plato belonged (Thompson 2000). Many believers would go hungry before they would violate these proscriptions. Even in a twenty-first century landscape populated by chain restaurants purveying uniform fare from Maine to Hawaii, what you eat still reveals who you are and where you're from. If you want to experience personally the cultural nature of food, walk into a New York deli and order corned beef on white with mayo. They will know you are not from the neighborhood.

Historically, people assumed that food "safety" and food "purity" were connected, and that relationship was built into early safety requirements. Today, food regulation applies a more limited definition of safety, based on the sciences of toxicology, microbiology, and nutrition. But people still yearn for purity and wholesomeness (Thompson 2000). Terms like "organic" and "free-range" address that yearning. They convey a sense of ritual cleanliness and foster a perception that foods bearing these labels are not just safe, but pure and wholesome as well (Belton 2001).

Psychologists who specialize in risk perception tell us that people are most fearful of those risks they perceive as unknown, uncontrollable, and potentially catastrophic (Slovic 1987). It should not be surprising that many consumers are uneasy about "genetically modified" and "transgenic" food. These products are produced by new and highly sophisticated technology that alters the most basic mechanisms of reproduction. They are perceived as artificial and unnatural, and as having the potential for catastrophic consequences.

Current Public Views of Biotechnology

It is difficult at this point to assess how much impact these concerns will have on consumer acceptance of genetically modified food in the future. In fact, Americans have been eating genetically altered food products for several years. Most of the corn and soybeans grown in the United States are genetically modified, and over half the foods in supermarkets have been produced, at least in part, using agricultural biotechnology. Eating the food does not necessarily signify acceptance, however. Genetically altered foods are ubiquitous but invisible. Consumers are simply unaware of their presence, because most of the corn and soybeans grown in the United States, including genetically modified varieties, is fed to animals and converted to meat and milk or used as ingredients in processed foods, such as corn sweeteners in soft drinks. The government doesn't require these foods to be labeled, and food processors have not been willing to provide label information that acknowledges the presence of GM ingredients.[1]

It's not surprising that surveys show that only about a third of respondents thought foods produced through biotechnology are currently on supermarket shelves. Few could accurately identify which foods were affected (International Food Information Council 2004; Pew Initiative on Food and Biotechnology 2004). Forty-four percent thought vegetables and 20 percent thought fruits were genetically modified, even though there are no such products on the market (International Food Information Council 2004).

While there is no market experience to determine how many Americans are comfortable with GM foods and how comfortable they are, there is a body of opinion research on what Americans know and how they feel about GM products, and whether they think they are safe and would choose to buy them if they knew more about them. In addition to the more general surveys done by Gallup and news organizations, there are some publicly available in-depth surveys that are useful in discerning consumer views about GM food products.

The International Food Information Council (IFIC), an industry-funded research organization, has conducted annual opinion surveys since 1999. Because they repeat some basic questions each year, their work provides one of the few sources of data on how views have changed over time. The Pew Initiative on Food and Biotechnology (PIFB), funded by the Pew Charitable Trusts to provide an objective source of credible information on agricultural biotechnology, began conducting polls on consumer views of biotech food in 2001 and has now extended its research to include focus groups that add depth to the survey data. A multistate, land-grant university research consortium conducted an in-depth survey in 2001, and Thomas Hoban, a sociologist from North Carolina State University who has examined public views on the subject both in the United States and around the world for fifteen years, has developed an impressive database.[2]

The goals and methods used by these and other researchers vary, and the differences may affect their results. The IFIC survey instrument is constructed to provide respondents favorable information about biotechnology before posing a question, and their questions usually mention potential benefits, but not potential risks. This approach can produce data that reflect a more positive view.[3]

Most research, however, has concluded that most consumers don't know much about food biotechnology. In September 2004, PIFB found 32 percent of respondents had heard a "great deal" or "some" about food biotechnology, a drop of fourteen points from a previous poll in 2001 (Pew Initiative on Food and Biotechnology 2004). IFIC, in 2004, reported 36 percent had heard a lot or some, a decline of seven points from January 2001 (International Food Information Council 2004).

Consumer views on the safety of genetically modified foods are evenly divided and haven't changed much over the past few years. An ABC News

TABLE 6-1

CONSUMER VIEWS: SAFETY OF GM FOODS (PERCENT)

PEW Initiative (2001)	Safe	Unsafe	Undecided
Are GM foods safe?	29	25	46

PEW Initiative (2004)	Safe	Unsafe	Undecided
Are GM foods safe?	30	27	43

Multistate, land-grant university consortium (2001)	Agree	Disagree	Undecided
Foods from plants that are genetically modified to add desirable food traits are safe.	28	25	47
Foods from animals that are genetically modified to add desirable food traits are safe.	17	39	43

SOURCE: Surveys by Pew Initiative on Food and Biotechnology (2001, 2004) and by a multistate land-grant university research team (S-276 Multi-State Research Team 2003). Data compiled by R. C. Wimberly.

poll taken in mid-July 2003 reported that 46 percent thought GM foods were safe, and the same percentage thought they were unsafe (ABC News 2003). The Pew Initiative and land-grant university consortium polls showed slightly more favorable results (see table 6-1). Pew reports that 30 percent of respondents in their most recent poll thought the products were safe. The 2001 land-grant consortium survey found that 28 percent thought products from GM plant products were safe (S-276 Multi-State Research Team 2003). However, only 17 percent said foods from genetically engineered animals were safe.

Consumer awareness of genetically modified ingredients in food and concern about their safety increase when there is a serious mistake or mishap that garners media attention. In late 2000, for example, StarLink, a variety of genetically modified corn that the Environmental Protection Agency had not approved for human consumption, was discovered in taco shells and other food products. There was enormous media coverage of the event. In 2001, when the Pew Initiative conducted its first opinion survey, it found that 44 percent of people polled had "heard a great deal" or "some" about biotechnology use in food production. While there have been subsequent incidents of unapproved products getting into the food supply, none has been as large or generated as much coverage. When Pew asked the same question in 2004, only 33 percent reported having heard a great deal or some.

A serious mistake such as StarLink also influences feelings about the safety of GM foods. An item on the 2001 Pew Initiative survey read, "The Centers for Disease Control (CDC) report found no evidence of allergic reaction attributable to StarLink. How concerned are you about the safety of eating genetically modified foods or about genetically modified foods in general?" Almost two-thirds of those surveyed, 65 percent, said they were concerned. Pew Initiative officials noted last year that "opinions formed during incidents such as the StarLink are lasting for some consumers." The group cited comments of a participant in a focus group conducted four years after StarLink. He claimed not to know much about GM foods but then stated, "I've been following the news. I can remember [there was] some stink storm when some genetically altered corn got loose in Minnesota into the general population. They're not approved across the board right now because [regulators] don't necessarily know how safe they are." (Pew Initiative on Food and Biotechnology 2004).

Pew Initiative researchers found some softening of overt opposition to GM foods between 2001 and 2004 but no change between 2003 and the last survey. The IFIC research reveals that the percentage of people who say they would be inclined to purchase potatoes or tomatoes genetically altered to reduce pesticide use dropped from 77 percent in 1997 to 66 percent in 2004 (see figure 6-1). Those inclined to purchase cooking oils altered to reduce saturated fat dropped from 57 percent in 1999 to 40 percent in 2004, a decline of seventeen percentage points in five years. Finally, IFIC asked whether people felt that biotechnology would provide benefits to them and their families within the next five years. In 1997, 78 percent responded favorably. By 2004 the favorable responses had dropped to 59 percent—still a majority, but showing a decline of nineteen percentage points in seven years (International Food Information Council 2004). This

FIGURE 6-1

CONSUMER WILLINGNESS TO BUY/EXPECTATION OF BENEFIT FROM
GM FOODS HAS DECLINED SINCE 1990s

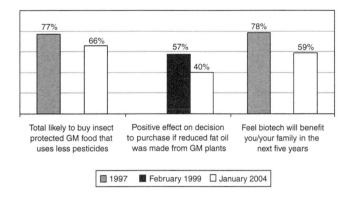

SOURCE: International Food Information Council (2004).

last response may represent both the biggest challenge and greatest opportunity for biotechnology supporters. Consumers have garnered little or no direct benefit from a decade of food biotechnology and now may be less sanguine about promises of future benefits.

Thomas Hoban has reported a similar trend: Consumer acceptance of insect-protected crops resulting from biotechnology dropped from 63 percent in 1992 to 51 percent in 2000. The proportion of respondents who found it unacceptable to use biotechnology to protect crops from insects rose from 18 percent in 1992 to 27 percent in 2000 (Hoban 2001).

Animal biotechnology—the creation of cloned and transgenic animals—arouses substantial opposition, with much of it grounded not in safety concerns, but in the moral and ethical acceptability of altering sentient beings. In its 2004 survey, IFIC added a question about animal biotechnology, prefacing it with a description of some potential beneficial uses of the technology, and stipulating FDA assurance of safety. Fifty-nine percent of respondents said they would be willing to purchase meat, milk, and eggs from transgenic animals, but 62 percent said they would be unwilling to purchase these products from cloned animals (see figure 6-2).

FIGURE 6-2

CONSUMER WILLINGNESS TO PURCHASE MEAT, MILK, AND EGGS FROM
CLONED AND TRANSGENIC ANIMALS

Question: *If the FDA determined that meat, milk, and eggs from* animals enhanced
through genetic engineering *or from* cloned animals *were safe, how likely would you
be to buy them?*

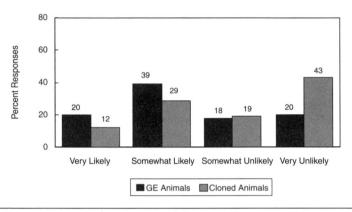

SOURCE: International Food Information Council (1997–2004).

Pew found a substantial level of discomfort with animal biotechnology. Asked about scientific research into the genetic modification of animals, 57 percent were opposed, with 46 percent saying they were strongly opposed (Pew Initiative on Food and Biotechnology 2004). The Gallup Organization each year asks consumers' view on the morality of various actions (see figure 6-3). In 2004, 64 percent said cloning animals was morally wrong, ranking only behind cloning humans (Gallup Organization 2004).

Virtually every poll taken reports that large majorities of consumers want food products derived from agricultural biotechnology to be labeled as such. While some researchers note that consumers don't volunteer an opinion that this labeling should be high priority (International Food Information Council 2004), there is little question that consumers want the information. The Pew Initiative found over 90 percent of consumers support the labeling of genetically modified foods and GM ingredients in processed foods (2004).

FIGURE 6-3
AMERICAN CONSUMERS' VIEWS THAT VARIOUS
ACTIONS ARE MORALLY WRONG

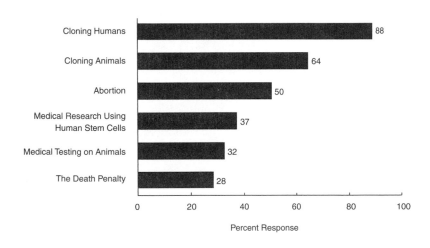

SOURCE: Gallup Organization (2004).

However, perhaps the most sobering information about public accept-ance of GM food comes from a 2003 survey conducted by the U.S. Depart-ment of Agriculture's Economic Research Service (ERS). Consumers were willing to pay an average of 14 percent more for food that appeared to be free of "genetically modified" ingredients. The ERS concluded, "Consumers' willingness to pay for food products decreases when the food label indicates that a food product is produced with the aid of modern biotechnology" (Tegene et al. 2003).

Finally, it is worth noting that women tend to be more concerned about risks from genetically altered foods and more likely to avoid purchasing them than men (ABC News 2003; Pew Initiative on Food and Biotechnology 2001). In general, women are more likely to perceive risk and to seek to avoid it (Slovik 1987). An ABC News survey from July 2003 is a typical example. The poll found that while 46 percent of all respondents thought gene-altered foods were unsafe, 54 percent of women surveyed held that view. While 55 percent of all respondents would avoid buying the foods if they were labeled, 62 percent of women would avoid them (ABC News 2003). Since women continue to be the primary food shoppers in most households (Food Marketing Institute 2003), their reaction is critically important to the success or failure of new food products, including those produced through agricultural biotechnology.

Why the Skepticism about GM Foods?

There are a number of possible explanations for public reluctance to embrace food biotechnology. These are summarized below.

Current biotechnology products were designed to benefit and appeal to food producers. In developing their products, biotechnology companies focused on seeds and crops, not foods. When technology producers say "consumers," they mean farmers, not supermarket shoppers. Their goal has been to lower production costs, not retail prices. Reducing the cost of producing corn is good for farmers, but it does not necessarily translate to lower prices at the supermarket. Moreover, none of the current generation of biotech crops offers consumers more nutritious or better-tasting food.

Since genetically engineered foods weren't designed to benefit consumers and have not done so, industry and government have scrambled retroactively to devise rationales for public acceptance. The arguments are unpersuasive and, sometimes, insulting. They urge people to accept currently available producer-oriented GM products because, someday, there will be new products that do have consumer benefits (Monsanto Company 2003). Although they've made that promise for over a decade now, there are no new consumer products on the market or even close to the end of the development pipeline.

Industry and government officials also argue that agricultural biotechnology will help feed the world and suggest that Americans and Europeans who resist the technology are contributing to world hunger (U.S. Trade Representative 2003). However, current food biotechnology products were not designed to meet the needs of Africans any more than they were designed to provide benefits to American consumers. None of the GM crops now on the market was originally developed to address either African agricultural constraints or African consumer preferences. Setting aside the unproven claim that biotechnology would reduce world hunger, it is unlikely that Americans would change deeply ingrained food habits and preferences in response to such exhortations.

One argument for GM products is both valid and likely to appeal to consumers. The technology has reduced reliance on some of the most toxic pesticides and appears to have the potential to reduce overall pesticide use. Given public concern about the human health and environmental risks created by pesticide residues, it is surprising that biotech supporters haven't emphasized this benefit. Is it because doing so might stimulate a serious discussion of the risks associated with pesticides, and why it might be better for human health and the environment to reduce their use? While this argument alone is not sufficient to persuade Americans to embrace food biotechnology, it has substantive merit, as well as some potential to increase public support.

No product or technology, especially food-related, comes to market in a vacuum. Acceptance of GM products is affected by the social and political context surrounding their introduction, and both government and industry have ignored current circumstances that arouse public concern. Food safety disasters over the past decade, particularly in Europe, have intensified food-related fears. Instantaneous communication has ensured that bad news will be heralded everywhere. No mistake by the food biotechnology industry or government goes unnoticed and, justifiably or not, each mistake, each misadventure, taints the entire biotech enterprise (Artuso 2003). News of the StarLink contamination spread around the world in a matter of minutes and, as the Pew Initiative poll showed, the resulting concern among consumers was still perceptible four years later (Pew Initiative on Food and Biotechnology 2004).

The fear and dislocation associated with globalization have stimulated antibusiness animus around the world, creating yet another cultural disconnect that affects public attitudes. Agricultural biotechnology is associated with increasing global corporate power. While our eating habits are deeply rooted in culture, the World Trade Organization is focused on economics. The SPS agreement limits the rights of governments to respond to cultural norms about food hazards to the point that cultural differences about food safety and purity can create unintentional, nontariff trade barriers.[4] Trade policy experts who argue that advancing free trade requires food be treated no differently than a bar of soap have to overcome the weight of centuries of tradition. It is too soon to know whether consumers will ultimately decide that the economic benefits of trade liberalization outweigh those cultural attachments (Echols 2001).

It's also too soon to know whether trade agreements can adjust to national political imperatives that relate to cultural attachments to food. The U.S. government has filed a WTO complaint against the European Union, citing both EU reluctance to approve GM products as safe and the union's insistence that all foods produced through agricultural biotechnology be labeled as such. The complaint ignores the reality that European consumers oppose genetically modified food. Members of parliament in the United Kingdom, France, and Germany are no more likely to tell their constituents that they must eat GM corn or wheat because it is good for American farmers and required by trade agreements than members of the U.S. Congress are to tell their constituents that they must buy Limburger cheese because it is good for Belgian farmers and consistent with free trade. As long as there are elections, parliamentarians and congressmen will put the home folks' interests ahead of world trade.

The food biotechnology industry suffers from self-inflicted wounds. Despite recurrent lapses, both industry and government continue to insist that the current regulatory system is adequate to protect human and environmental health and safety. Government risk managers and industry scientists insist that they can deploy the technology free from mistakes, accidents, and unintended negative consequences. They assured the public that corn unapproved for human consumption would never get into the food supply, that pharmaceutical corn would never contaminate food corn, and that neither experimental transgenic animals nor their offspring would ever get out of the laboratory and into the packing house. But every one of these things has already happened. StarLink corn was found in a variety of human food products. ProdiGene, a drug corn, was mixed with food corn. The University of Illinois at Champaign-Urbana allowed "no take" pigs (nontransgenic offspring of bioengineered pigs) to be sent to a plant that slaughters animals for food. Having had their claims of infallible regulation proved wrong, biotechnology producers and government then argued that the failures didn't matter because no one got sick. These incidents have passed from the front pages and evening news but, as the Pew Initiative's focus groups have demonstrated, the memories linger on in the public consciousness and threaten public confidence in government safety efforts.

The adamant refusal to label genetically modified products almost surely contributes to public suspicion that there is something wrong with them. It is human nature to suspect that people who act covertly have something to hide. The opposition to mandatory labeling, despite a plethora of polls showing that consumers overwhelmingly want this information, impairs the credibility not just of biotech producers, but of food processors and government regulators as well.

New Generation of Products May Exacerbate Consumer Discomfort

While genetically modified corn and soybeans have been imperceptible to retail consumers, three new GM products are likely to be more visible and more controversial.

Genetically Modified Wheat Used in Bread, Cereal, and Pasta. Roundup Ready wheat may be no different scientifically from Roundup Ready soybeans, but it has a different cultural impact and is likely to be more detectable and raise more concern.

Wheat is a prominent human staple, the "staff of life." It is the primary ingredient in our daily bread, our children's cereal, the communion wafer, and Passover matzo. Roundup Ready wheat may not be listed on ingredient labels, but its presence won't be a secret for long. It will be easy for opponents to test breads and cereals and tell consumers which ones contain it. Furthermore, American wheat growers export 50 percent of their production and command top prices in world markets. Monsanto continues to urge wheat farmers to adopt their biotech product despite the fact that many of those foreign markets reject GM food.

Corn Modified to Produce Pharmaceutical and Industrial Substances. Plant-made pharmaceuticals (PMPs) and plant-made industrial products

(PMIPs) carry the risk that drugs will turn up in breakfast cornflakes, cooking oils, and soft drinks. Food processors and retailers, who up to this point have been supportive of the new technology, are now voicing serious concerns about these products.

The National Food Processors Association (NFPA) filed comments with the USDA on pharmaceutical plants, stating that

> given a voice . . . NFPA would not have supported the use of food plants for the production of plant made pharmaceuticals (PMPs). The risk and impact of contamination to the food supply is simply too great, as the food industry learned through experiences with the commodity crop StarLink corn. . . . NFPA strongly opposes the use of food crops to produce PMPs commercially without effective controls and procedures that ensure against any contamination of the food supply (National Food Processors Association 2003).

NFPA demanded "100 percent protection from PMP contamination," stating that anything less "would undermine consumer confidence at home and jeopardize international trade" (National Food Processors Association 2003).

Corn producers, their friends in Congress, and government regulators insist it is possible to grow pharmaceutical corn in the middle of the Corn Belt with no risk that the altered product will find its way into the human food supply. Such conviction is a triumph of hope over reality. It requires extraordinary confidence in the competence and integrity of the people growing, shipping, and milling these products. It also speaks of a faith that government regulation will work perfectly—a faith that farmers and food processors usually do not exhibit.

Transgenic and Cloned Animals. The manufacture of cloned and genetically altered animals and the issue of whether they will become part of our food supply raise moral, ethical, and cultural questions that science cannot resolve, and neither industry nor government can ignore. The FDA is compiling scientific data on the safety of cloned animals and has asked producers voluntarily to withhold their meat and milk from the food supply until the agency completes its work. However, government is not examining the qualitative, nonscientific issues raised by altering sentient beings, issues that have not been a major point of concern in the discussion of biotech crops, but are likely to arise with regard to animal biotechnology.

As a National Academy of Sciences committee noted in its 2002 report on animal biotechnology, the creation of transgenic and cloned animals raises moral and ethical issues that have little to do with science but are important to the public (National Research Council/National Academy of Sciences 2002). We relate differently to animals. When the first biotech corn hit the market, no one thought to name it Bt Betty and flash photos on television. The birth of Dolly, the first cloned sheep, touched off a media frenzy. Her placid face stared out from newspapers and television sets around the world. Many people who have never cared one whit about genetically modified organisms in plants cringe at the thought of creating a salmon that grows five times as fast as its natural counterpart, altering sheep to produce unshrinkable wool, or cloning pets so Fido can live forever. Opinion research shows strong concern about using biotechnology to create transgenic animals, and opposition to animal cloning (Pew Initiative on Food and Biotechnology 2004; International Food Information Council 2004; Hoban 2001).

To put it in practical rather than scientific terms, creating transgenic and cloned animals and eating meat and milk from them evokes a common response—"Yuck."

The Current Regulatory System Undermines Support for Food Biotechnology

While tradition continues to be an important determinant of food choices, modern consumers, far removed from any role in food production or processing, increasingly rely on the food industry and regulators to ensure food safety (Echols 2001). Rigorous and credible regulation not only reduces the possibility that unsafe products will harm the public or the environment, it increases public confidence in and acceptance of new technologies. The public must trust that risk managers and the risk management system are competent, honest, and devoted to protecting the public interest. Trust is the defining element in determining public acceptance of a given risk. It is a valuable and fragile commodity, hard for government risk managers to build and extremely easy for them to lose (Slovic 1987). If the public senses that risk managers are not caring or competent, their concern rises and may result in the adoption of overly restrictive remedial action (Powell 2000).

Different groups in society view risk differently. Scientists and consumers frequently differ about risk. Scientists are more inclined to consider only data from the hard sciences, while the public level of concern is more likely to respond to whether the threat is potentially catastrophic, exotic, hidden, or unequally applied (Slovik 1987). Industry and government tend to address any controversy over regulation by stating that it is based on "sound science." In recent debates, "sound science" has become a mantra, apparently with the expectation that its ritual chanting will end all further debate. In fact, if the term were ever meaningful, constant repetition has robbed it of any consequence.

For a number of reasons, citizens are not always inclined to accept the assertion that a position reflects "sound science" or is "science-based" as dispositive. People see that scientific orthodoxy changes regularly, and that today's scientific certitude is often tomorrow's error. Newspapers and television regularly report scientific conflicts and reversals. It is possible to compile a considerable file, for instance, on the single subject of conflicting scientific findings on diet and obesity.

Although business and government assert that decisions are based only on scientific "fact," most people understand that science is not value free. Scientists examine the same data and come to different conclusions based on the values and interests they bring to the process. Scientific studies frequently reflect positions consistent with the interests of those who funded the work. The National Academy of Sciences Committee on Animal Biotechnology explained the play of values in science and policy, noting that "while values shouldn't alter scientific analysis, they inevitably and properly influence the procedural framework in which scientists address questions that have regulatory consequences and they shape public policy responses to science-based health and environmental concerns" (National Research Council/National Academy of Sciences 2002). Regulations might have more credibility if we acknowledged that competing sets of values and interests influence regulatory actions.

In most cases, neither science nor law is so narrow that it dictates one particular course of action. Risk managers usually have plenty of leeway to choose between a minimally restrictive regulatory regime that might increase the risk to health and safety and a more restrictive approach that might stifle innovation. Policy decisions require choices between these competing values and interests. That's why making them is never as simple as the assertion that one course of action reflects "sound science" and another does not. It's why risk management is hard work.

In the 1986 Coordinated Framework for Agricultural Biotechnology, the U.S. government chose a regulatory path intended to encourage the rapid development and introduction of food biotechnology, allowing American companies to take advantage of their head start in this field (U.S. Office of Science and Technology Policy 1986). The framework was based primarily on the values and interests of biotechnology companies and farmers, in part because the public was largely unaware of the technology or that a decision-making process was underway. Now that the public is aware and concerned, the weaknesses of the framework are apparent, and they contribute to public disquiet in the following ways:

The current regulatory system is convoluted, occasionally opaque, and, in some areas, inadequate to ensure safety. Most of the laws governing biotechnology were written before the field was developed or even dreamed of. Three agencies and ten statutes overlap and, occasionally, leave serious gaps in protection. There are no clear, logical lines of authority, making it difficult to assign responsibility when something goes wrong. The system is unintelligible to the public. As figure 6-4 shows, the regulatory structure for food biotechnology looks like a Rube Goldberg contraption, inviting public ridicule.

The FDA, the nation's primary food safety agency, does not examine and declare plant products safe for human consumption before they are allowed on the market. Although it is reasonable to assume most Americans believe that ensuring the safety of a new food product for human consumption is the most basic requirement of a food safety regulatory system, the current scheme governing development and use of genetically modified products provides more safeguards for plant protection and environmental safety than for human health.

FIGURE 6-4

GOVERNMENTAL REGULATION OF GENETICALLY MODIFIED ORGANISMS IN FOOD

Products derived from transgenic organisms are regulated according to their attributes and intended use.

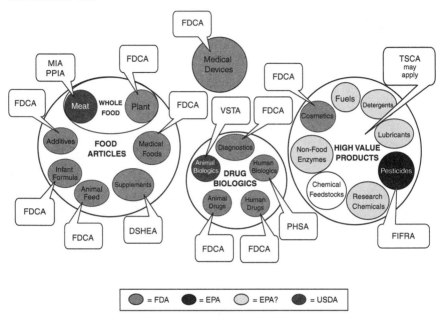

SOURCE: Fish and Rudenko (2001). Reprinted with permission.
NOTE: FDCA = Food, Drug and Cosmetic Act; DSHEA = Dietary Supplement Health and Education Act; PHSA = Public Health Service Act; FIFRA = Federal Insecticide, Fungicide and Rodenticide Act; TSCA = Toxic Substances Control Act; MIA = Meat Inspection Act; PPIA = Poultry Products Inspection Act; VSTA = Virus, Serum Toxin Act.

It is illegal to market a genetically modified plant without first applying for and receiving approval from the USDA certifying that the plant will not harm other plants (U.S. Department of Agriculture 1997). It is also illegal to market a genetically modified, pest-protected plant without first applying for and receiving approval from the EPA certifying that the plant will not harm the environment (U.S. Environmental Protection Agency 1994). However, there is no legal requirement for a company to demonstrate that a genetically modified plant will not harm humans. In protecting human health, the U.S. Food and Drug Administration employs a voluntary consultation system. Producers of genetically altered food products are expected, but not required, to consult with the FDA before marketing the product. The FDA acknowledges that the agency does not conduct a comprehensive scientific review of the data generated by the developer. It does not make a positive determination that a new food biotechnology product is safe, nor affirmatively approve its sale before it goes to market (U.S. Food and Drug Administration 1997).

At the end of the voluntary review process, the FDA sends a letter to the biotechnology company stating that the company has concluded its product is as safe as other varieties of the product currently on the market. The letter reminds the company that it has a continued responsibility to ensure the foods it markets are safe (Rulis 1995). Receipt of the letter signals to the company that the FDA will not challenge the marketing of the product. This policy is grounded on an assumption that no company will go to market without consulting the FDA; no company will ever make a mistake in its assessment or fail to provide all relevant information; no company will ever be driven by pride or economic incentive to overstate its case.

The current system is not adequate to address issues raised by the more sophisticated products that are in the pipeline. It appears that companies have so far engaged in the voluntary process at least to some extent, and that no human health problems have arisen from foods currently on the market. However, a recent study shows that, contrary to assertions by both the FDA and the industry, companies have not always provided the FDA with all of the information sought by regulatory officials and, occasionally, have provided incorrect information (Jaffe and Gurian-Sherman 2003). The Pew Initiative reports that both its focus groups and surveys revealed consumers were surprised and concerned that the FDA does not have authority to examine genetically engineered products and formally approve their safety before the products are sold. Respondents indicated they would be more comfortable with the technology if there were such a formal premarket process, overseen by the nation's primary food safety agency (Pew Initiative on Food and Biotechnology 2002, 2004).

The process for approving transgenic and cloned animals is cloaked in secrecy. With the possible exception of national security issues, Americans insist on their right to know what their government is doing and to have a voice in how it is done. The FDA regulates transgenic animals under the New Animal Drug Provisions of the Food, Drug and Cosmetic Act (U.S. Food and Drug Administration 1997). The safety standard is rigorous. The FDA must find that the product is both safe and effective before it is sold.

However, the entire process for approving these animals is secret, with no public participation. The FDA does not announce that a new application has been filed, nor does it make any information about the application available to the public, receive comments from the public, issue a public statement to justify its action approving the product, or make the public aware of the data on which the decision was based. Given Americans' discomfort with transgenic animals, secrecy cannot advance either the public or private interests. With regard to the regulation of cloned animals, the FDA will probably find that, because they are by definition the same as existing animals, there is no reason to regulate them or the meat and milk they produce for human food.

There is no mechanism for explicitly addressing issues that trouble the public. The public focuses its concerns not just on the health and environmental impacts of biotechnology, but also on questions like, "Should we be doing this?" "Is it ethical to alter sentient beings?" and "Why should we take any risk for these products?"

FDA officials insist that they look only at science and do not address social, cultural, and ethical issues. The agency does, however, address more than science. By law, it considers the economic impact of its decisions. The agency is also influenced by strong public and congressional reaction to its policies and by negative media attention. It will often respond by moving slowly on decisions officials know will be controversial, requiring additional studies to try to reassure the public or insisting that products be labeled.

However, whatever the underlying imperative, the FDA will posit a scientific rationale for the action. In response to the increasing opposition to food biotechnology and the Coordinated Framework, the FDA in 2001 proposed a new regulation that would have made the voluntary consultation process mandatory. This was an effort to reduce public discomfort but avoid any action that would upset the industry, require substantial additional resources, or necessitate changing the law.

But the FDA refuses to address explicitly the greater public concerns about food biotechnology. The refusal doesn't make the concerns go away.

It simply ensures that nonscientific issues—social, ethical, and religious concerns—will be brought to the agency in the guise of concerns about science. And it ensures that public trust in the risk management system and the technology will be diminished. There are good reasons to be cautious about government regulatory agencies making policy based on moral and ethical issues, but government can and should find appropriate ways to address these concerns.

Americans want to know what is in their food. As noted earlier, every poll shows that the public wants genetically modified foods to be labeled as such. The industry's refusal to do so and the FDA's acquiescence in concealing their presence further undermine public trust in both the products and the system. Both industry and government seem oblivious to the possibility that the continuing lack of enthusiasm for food biotechnology may reflect an increasing awareness of the technology and a nagging lack of trust in the system and the risk managers to control risks and protect the public interest.

In recent speeches and testimony, the Bush administration has adamantly supported the current system and denigrated public concerns. FDA officials signaled that they believe the agency's current regulatory system is sufficient to deal with food safety issues. They've suggested they will allow companies producing PMPs to declare that at least some of the drugs grown in corn are generally recognized as safe (Flamm 2003). This means that if drug corn were to get into the human food supply, it would not be necessary to recall food products made from the corn. Both the industry and the FDA could assure consumers that there is no reason to be concerned about the addition of a little pharmaceutical material to their cornflakes, and farmers and food processors would be protected against liability suits. The FDA has virtually abandoned even the small improvements offered by regulatory changes for plant products proposed in 2001 and indicated they see no reason to make the process for approving transgenic animals more transparent (Crawford 2003a, 2003b).

It is understandable that the food biotechnology industry fears a more rigorous, transparent, and comprehensible regulatory system will increase the time and costs of bringing products to market. But at least one study suggests that the economic benefits would outweigh any additional financial burden. Anthony Artuso (2003), a professor of agricultural, food, and resource economics at Rutgers University, argues that the higher the perceived risk of harm, the greater the degree of regulation needed to maintain or restore consumer confidence in a food product. Adverse events—mistakes, accidents, mishaps—are less likely to occur when regulation is rigorous. Since any adverse event involving any product of food biotechnology causes consumers to react negatively to all GM products, it benefits all producers to avoid incidents and the resultant negative publicity. Artuso concludes that regulatory requirements going beyond those based solely on scientific risk assessment are economically efficient in the long run, even if initial product development is more costly. His theory is supported by focus group research in which participants agreed they would be more comfortable and supportive of food biotechnology if regulation were more protective (Pew Initiative on Food and Biotechnology 2003).

Practical Steps to Improve Prospects for Public Acceptance of Food Biotechnology

Reasonable people should be able to find a way to enjoy the potential benefits of food biotechnology while avoiding unintended negative consequences. If government and industry were willing to take steps to address public concerns and remedy weaknesses in the regulatory system, most Americans could probably be persuaded to embrace the new technology. There are a number of measures the federal government and the biotechnology producers and food industry should consider taking.

Strengthen the regulatory system. The government should strengthen the food regulatory system by providing clear lines of regulatory authority; giving the FDA sufficient power and resources to examine thoroughly and approve the safety of plant products, including nonfood products that are to be produced in food plants, before they enter the market; and making the animal approval system more transparent. These changes would require congressional action, which the industry opposes on grounds it might not be able to control the legislative process in the same way it influences regulatory actions. At present, that risk is minimized by the fact that both the Bush administration and congressional leaders strongly support food biotechnology. Making necessary reforms now, before there is a serious mishap, will help protect public health and the industry's future.

Address directly and immediately the moral and ethical issues raised by genetic manipulation of animals. The moral and ethical concerns surrounding cloned and transgenic animals deserve explicit recognition and a thorough public discussion at the highest level of government. The president might direct his Commission on Bioethics to consider the issues, or the president or Congress might convene an ad hoc group of ethicists, religious leaders, and social scientists to examine the implications for society of altering sentient beings and perhaps develop a framework for determining the kinds of activities that are acceptable to the public.

The biotechnology and food industries should not only support but also actively urge the federal government to act on these issues. In addition, given the potential threat to markets for foods produced by the United States, biotechnology companies might reconsider their rush to market Roundup Ready wheat and to grow pharmaceuticals and industrial products in corn, at least in the Corn Belt states. While nonfood crops may not be as desirable for producing these products, they can be adapted and used. It may be cheaper and faster than overcoming public outrage after a contamination event.

Shift focus to view retail consumers as the ultimate arbiters of whether food biotechnology will meet its potential. Biotechnology producers should take a crash course in how to think like consumer product companies. This would probably lead them to concentrate on bringing to market some long-promised products with direct consumer benefits. Offering retail consumers a product they are eager to buy could change the context of the debate. Since companies would want to identify the desirable traits of these products, they'd want to label them, providing information consumers want.

Food processors and retailers might also consider reevaluating their role in this debate. For over ten years, they have dutifully defended GM products and promoted the biotechnology industry's campaign for public acceptance and against effective regulation, even after the StarLink and ProdiGene controversies. Flouting this loyal support, the agricultural biotechnology companies are ignoring pleas from processors and retailers to forgo growing drugs in corn. As good businesspeople, food processors and retailers should protect themselves, their products, and their customers.

Both biotechnology companies and U.S. government officials might try a new approach to dealing with criticism of the current system. They have tended to respond to policy concerns as attacks on the very existence of the technology, either demonizing or dismissing anyone who questions current policy.

While there are groups that have and will continue to oppose bitterly any use of agricultural biotechnology, others are willing to work with the industry to address legitimate public concerns. The Pew Charitable Trusts made a valiant attempt to advance a consensus approach to a new regulatory policy. Their Initiative on Food and Biotechnology established a stakeholder forum, bringing together a group of eighteen people who represented virtually all of the various interests affected by food biotechnology. The group was not able to reach a consensus in the short term, but participants learned they could agree on problem areas and the elements of a rigorous and robust regulatory system.

Just as importantly, they learned they could work together, learn from each other, and come to respect and even like one another. Most members of the forum agreed that targeted changes in current statutes and regulations would provide substantially increased human health and environmental protection, and that the changes would lead to increased public comfort with the technology. Most were confident that a broad coalition of environmentalists, consumers, farmers, and biotechnology and food companies working toward that goal would greatly enhance chances of success and minimize risks to the industry.

There is still time for the biotech companies and their farm-group allies to reach out to other interested parties and seek compromise before the inevitable crisis arises. Science does not require food biotechnology and food processing companies or the U.S. government to take any of these steps. Common sense and self-preservation do.

Notes

1. Some processors have used their labels to proclaim the absence of genetically modified ingredients, but the government has imposed strict limitations on what can be stated.

2. Additional in-depth studies have been conducted over a period of years by a consortium of researchers, funded by the National Science Foundation. Their research compares U.S. and European views (see Priest 2000). Cook College of Rutgers University has completed two major surveys, and additional publicly available studies are funded by groups that actively support or oppose food biotechnology.

3. The American Association for Public Opinion Research's "Best Practices for Survey and Public Opinion Research" (2002) notes that the manner in which questions are asked, as well as the specific response categories provided, can greatly affect the results of a survey: "Question wording should be carefully examined for special sensitivity or bias."

4. SPS is the WTO Agreement on Sanitary and Phytosanitary Measures. It grants WTO members the right to protect humans, plant, or animal life, or health affected by international trade, but only to the extent that the measures adopted have minimal negative effects on trade. The primary objective of the SPS agreement is to promote international trade in agricultural goods and commodities.

References

ABC News. 2003. Poll: Modified Foods Give Consumers Pause. http://abcnews.go.com/sections/business/Living/poll030715_modifiedfood.html (accessed March 29, 2005).

American Association for Public Opinion Research. 2002. Standards and Best Practices for Survey and Public Opinion Research. http://www.aapor.org/default.asp?page=survey_methods/standards_and_best_practices/best_practices_for_survey_and_public_opinion_research (accessed March 28, 2005).

Artuso, A. 2003. Risk Perceptions, Endogenous Demand and Regulation of Agricultural Biotechnology. *Food Policy* 28:131–45.

Belton, P. 2001. Chance, Risk, Uncertainty and Food. *Trends in Food Science and Technology* 12 (1): 32–35.

Crawford, Lester. 2003a. Remarks at the American Enterprise Institute. Washington, D.C. June 12.

———. 2003b. Testimony before the Subcommittee on Conservation, Credit, Rural Development and Research. U.S. House of Representatives. Committee on Agriculture. *Hearing to Review Biotechnology in Agriculture.* 108th Cong., 1st Sess. June 17.

Echols, M. 2001. *Food Safety and the WTO: The Interplay of Culture, Science and Technology.* The Hague: Kluwer Law International.

Fish, A., and L. Rudenko. 2001. Guide to U.S. Regulation of Genetically Modified Food and Agricultural Products. Prepared for the Pew Initiative on Food and Biotechnology. September 7.

Flamm, E. 2003. Comments at BIO convention. Washington, D.C. June 25. Reported in *Food Chemical News.* June 30, 16.

Food Marketing Institute. 2003. Trends in Food Marketing 2003. Washington, D.C.: Food Marketing Institute.

Gallup Organization. 2004. Gallup Poll. May 2–4. Data provided by the Roper Center for Public Opinion Research, University of Connecticut.

Hoban, T. J. 2001. American Consumers' Awareness and Acceptance of Biotechnology. Presented at NABC Winter/Spring council meeting. March 6. www4.ncsu.edu/.../articles/Amerpercent20Consumerspercent20Awarenesspercent20andpercent20Acceptancepercent20ofpercent20biotechnology.pdf (accessed May 2003).

International Food Information Council (IFIC). 2004. US Consumer Attitudes toward Food Biotechnology. *Cogent Research.* January. http://ific.org/pdf/2003BiotechSurvey.pdf (accessed March 2005).

Jaffe, G., and D. Gurian-Sherman. 2003. Plugging Holes in the Biotech Safety Net. *CSPI.* January 7.

Monsanto Company. 2003. Biotech Basics: The Benefits of Biotechnology. http://www.biotechknowledge.monsanto.com/biotech/bbasics.nsf/benefits.html (accessed April 25, 2005).

National Food Processors Association. 2003. Field Testing of Plants Engineered to Produce Pharmaceuticals and Industrial Compounds. Comments re: docket no. 03-031-1. *The Federal Register* 68 (March 10): 11337.

National Research Council/National Academy of Sciences. 2002. *Animal Biotechnology: Science-Based Concerns.* Washington, D.C.: National Academy Press.

Pew Initiative on Food and Biotechnology. 2001. Poll conducted by the Mellman Group and Public Opinion Strategies. January 22–28.

———. 2002. Unpublished focus group report on public attitudes toward genetically modified foods and agricultural biotechnology.

———. 2004. Overview of Findings, 2004 Focus Groups and Poll. http://Pew Initiativeagbiotech.org/research/2004update/ (accessed March 28, 2004).

Powell, D. 2000. Food Safety and the Consumer—Perils of Poor Risk Communication. *Canadian Journal of Animal Science* 80:393–404.

Priest, S. Hornig. 2000. US Public Opinion Divided over Biotechnology. *Nature Biotechnology* 18 (September): 939–42.

Rozin, P. 2000. The Psychology of Food and Food Choice. In *The Cambridge World History of Food*, ed. K. F. Kiple and K. C. Ornelas. Cambridge: Cambridge University Press.

Rulis, A. (Acting director, Office of Pre-market Approval, Center for Food Safety and Applied Nutrition, U.S. Department of Health and Human Services, Food and Drug Administration). 1995. Letter to Diane Re (Regulatory Affairs, Agricultural Group of Monsanto). January 27.

S-276 Multi-State Research Team. 2003. The Globalization of Food and How Americans Feel about It. *Southern Perspectives* 6 (2). http://srdc.msstate.edu/publications/wintersp03.pdf (accessed March 28, 2005).

Slovic, P. 1987. The Perception of Risk. *Science* 236:280–85.

Tegene, Abebayehu, Wally Huffman, Matthew Rousu, and Jason Shogren. 2003. *The Effects of Information on Consumer Demand for Biotech Foods: Evidence from Experimental Auctions.* Bulletin No. TB1903. U.S. Department of Agriculture. Economic Research Service. April.

Thompson, P. 2000. Incorporating Ethical Considerations. In *Encyclopedia of Ethical, Legal and Policy Issues in Biotechnology*, ed. T. H. Murray and M. J. Mehlman. New York: John Wiley & Sons.

U.S. Department of Agriculture. Animal and Plant Inspection Service. 1997. Regulations under the Federal Plant Protection Act. 7 CFR §340.3(b).

U.S. Environmental Protection Agency. 1994. Regulations under the Federal Insecticide, Fungicide and Rodenticide Act for Plant-Incorporated Protectants. 40 CFR Parts 152 and 174. http://www.epa.gov/fedrgstr/EPAPEST/1994/November/Day-23/pr-53.html (accessed March 28, 2005).

U.S. Food and Drug Administration. Center for Food Safety and Applied Nutrition. 1997. Guidance on Consultation Procedures: Foods Derived from New Plant Varieties. October. http://www.cfsan.fda.gov/~lrd/consulpr.html (accessed April 26, 2005).

U.S. Office of Science and Technology Policy. 1986. Coordinated Framework for Regulation of Biotechnology Products. *The Federal Register* 51:23302.

U.S. Trade Representative, Office of. 2003. Press release. May 13. http://www/ustr.gov/new/biotech.htm (accessed June 2003).

Visser, M. 1986. *Much Depends on Dinner*. New York: Grove Press.

PART III

Solutions

Introduction

The process of taking research on new biotech products from a promising early stage to commercial reality is fraught with challenges, in part because of harsh public perceptions that shadow the industry. Today these issues are filtered for the public through the dark lens of the protest industry, which has been successful, particularly in Europe, in exploiting fear of the future and a general wariness about science and technology. How can this dynamic be changed so the debate can be opened to a wider array of informed perspectives?

The potential risks and rewards of agricultural biotechnology must be discussed in a global context. Our final three essays do just that. In "Deconstructing the Agricultural Biotechnology Protest Industry," Jay Byrne, principal in the Internet-savvy public relations firm v-Fluence Interactive, provides a valuable service in his analysis of the flow of money and influence in the antibiotech protest movement. Byrne brings great expertise to this subject. He has spent twenty years managing risk for government agencies and corporations, including Monsanto, focusing especially on biotechnology issues. Byrne cofounded v-Fluence Interactive, which has a cogent understanding of the growing influence of the Internet and its use by NGOs to sway public discourse.

In "'Functional Foods' and Biopharmaceuticals: The Next Generation of the GM Revolution," Martina Newell-McGloughlin, the director of the University of California Systemwide Biotechnology Research and Education Program, offers a roadmap for how promising biopharmaceutical products might avoid the political traps that have bedeviled the biotechnology industry to date. Newell-McGloughlin is also the codirector of the National Institutes of Health training program in biomolecular technology, and a member of the World Trade Organization Genomics Panel on Technology, the International Food Information Council Expert Panel, and the United Nations Technology Discussion Panel on Sustainable Agriculture. She is hopeful that if the public can be educated to the vast potential of a new breed of potentially lifesaving drugs developed through genetic modification, the resistance to the biotechnological revolution will fade.

Patrick Moore's "Challenging the Misinformation Campaign of Anti-biotechnology Environmentalists" is a fitting essay to conclude *Let Them Eat Precaution*. Moore, a founder of Greenpeace, has broken with the NGO over the issue of biotechnology. He knows the protest industry from the inside. Moore points to his own journey from environmental pessimist to eloquent spokesperson for constructive change as a sign that the future of agricultural biotechnology may yet be brighter than the recent past and skeptics might suggest. "There is a middle road based on science and logic," Moore writes, "sometimes referred to as common sense."

7

Deconstructing the Agricultural Biotechnology Protest Industry

Jay Byrne

Agricultural biotechnology has provoked a media and marketing debate of a pitch and vigor rarely seen, even by the standards of controversies over other technological innovations. The intensity and confrontational tone of the debate can be traced to a new kind of professional activist, one who combines money and marketing with the growing influence of the Internet to sway public opinion and public acceptance. Often portrayed as a grassroots, shoestring movement, the groups that oppose biotechnology are, more accurately, part of a much larger coalition of social activists, environmental nonprofits, and social-investment organizations, backed by a reservoir of funding from special-interest foundations.

Biotechnology is only the latest high-profile issue targeted by activist groups. This protest coalition is exploiting current concerns over globalization and leveraging complex issues like biotechnology for purposes that are not openly disclosed or easily recognized by the public. The three key forces behind this movement are *money*, mostly from "progressive" foundations; *marketing*, drawing on the combined resources of the anti-science wing of the environmentalist movement, the organic and natural products industry, and similar opportunistic feeders such as the "socially responsible" investment community; and the *Internet*, which ties the coalition together and provides a way to reach donors, spread campaign messages, market to consumers, and, most key, influence the media, who shape public perceptions.

Money

Biotechnology—genetic engineering or genetic modification (GM), in preferred activist parlance—has become one of the protest industry's top causes and an effective spur for fundraising. The antibiotech industry is not, as popular myth would have it, funded by small contributions from many individuals. v-Fluence, a public affairs and issues management agency with several biotechnology-related clients, undertook a review of publicly available federal income tax returns; corporate, foundation, and nonprofit annual reports; and research reports from watchdog groups like the Center on Consumer Freedom. As illustrated in table 7-1, since 1994, more than $750 million in philanthropic support has been directed to antibiotechnology groups such as Greenpeace, Friends of the Earth, Institute for Agriculture and Trade Policy, the Organic Consumers Association, and hundreds of other activist organizations targeting agricultural biotechnology. The amounts listed pertain only to leading protest groups and initiatives, and do not reflect all monies donated to antibiotechnology campaigns.

More than forty major organizations based in the United States participate in significant antibiotechnology events, publicly lobby against biotechnology, oversee websites dedicated to antibiotechnology campaigns, and/or play leadership roles in funding the antibiotechnology movement. Table 7-2 lists the financial expenditures of the top activist groups targeting biotechnology (of which spending on biotechnology protests represents an undetermined fraction). Total spending exceeds $600 million annually.

What do these groups have in common, other than a shared, unflagging opposition to biotechnology? Almost all of them support (and in turn receive support from) the organic/natural products and corporate social responsibility/ socially responsible investment (CSR/SRI) industries. These and hundreds of other foundations, corporate donors, and a small group of wealthy individual donors have heavily financed antibiotechnology activists and their campaigns during the past ten years. Who gets the money they distribute?

Many of these organizations and their donors may appear to be acting independently, and individually their budgets may not seem excessive. They are generally not scrutinized by the media, which portray them as diverse and independent grassroots organizations. However, an examination

TABLE 7-1

PHILANTHROPIC SUPPORT FOR ANTIBIOTECHNOLOGY ACTIVIST GROUPS

Foundation name	Sample amounts funded to antibiotechnology activist groups (in US $)
Andrew Mellon Foundation	2,380,000
Arca Foundation	735,000
Ben & Jerry's Foundation	207,500
Body Shop Foundation	40,000
Carnegie Corporation of N.Y.	3,512,000
Charles S. Mott Foundation	10,173,040
David & Lucile Packard Foundation	8,579,397
Doris Duke Charitable Foundation	635,500
Edward J. Noble Foundation	775,000
Flora Family Foundation	200,881
Florence & John Schumann Foundation	6,782,500
Ford Foundation	39,978,020
Foundation for Deep Ecology	4,158,800
Gaia Fund	278,300
Gap Foundation	643,000
Geraldine R. Dodge Foundation	2,636,000
Henry Luce Foundation	670,000
HKH Foundation	670,000
Irene Diamond Fund	375,000
Jennifer Altman Foundation	795,000
John D. & Catherine T. MacArthur Foundation	11,906,500
John Merck Fund	4,673,800
Joyce Foundation	14,583,000
McKnight Foundation	2,795,800
Overbrook Foundation	1,689,500
Patagonia Fund & Patagonia, Inc.	106,500
Pew Charitable Trusts	130,996,900
Ploughshares Fund	1,257,800
Richard & Rhoda Goldman Fund	7,485,000
Richard K. Mellon Foundation	4,630,000
Robert Wood Johnson Foundation	574,700
Rockefeller Brothers Fund	7,321,000
Rockefeller Family Fund	655,000
Rockefeller Foundation	3,375,000
Solidago Foundation	456,000
Tides Foundation & Center	1,500,000
Turner Foundation	8,282,000
W. K. Kellogg Foundation	2,815,000
Wallace A. Gerbode Foundation	1,106,000
Wallace Genetic Fund	1,905,000
Wallace Global Fund	1,596,000

SOURCE: Publicly filed IRS 990 forms, available at www.guidestar.org.

TABLE 7-2
FINANCIAL EXPENDITURES OF LEADING ANTIBIOTECHNOLOGY
ACTIVIST GROUPS, 2002

Activist group	Amount (US $)
American Humane Association	10,366,805
CAL PIRG	466,866
Center for a New American Dream	1,595,554
Center for Food Safety	705,402
Consumers Union U.S.	162,992,456
Council for Responsible Genetics	345,710
Earth Island Institute	5,271,042
Environmental Defense	41,339,781
Environmental Media Services	2,148,056
Environmental Working Group	2,302,243
Farm Sanctuary	2,783,945
Foundation on Economic Trends	438,815
Friends of the Earth U.S.	4,644,563
Global Resource Action Center	2,837,193
Greenpeace International	112,332,000
Greenpeace USA	9,809,744
Humane Society of the United States	67,272,795
INFACT	1,084,735
Institute for Agriculture and Trade Policy	3,806,614
Institute for Food & Development Policy	1,297,388
Institute for Social Ecology	297,018
Int'l Center for Technology Assessment	566,190
Interfaith Center on Corporate Responsibility	1,333,636
MA PIRG	582,089
Mothers for Natural Law (2001)	110,000
National Family Farm Coalition (1999)	152,701
Natural Resources Defense Council	43,370,521
NY PIRG	4,802,746
Oldways Preservation (Chefs Collaborative)	883,539
Organic Consumers Association	1,249,727
Pesticide Action Network North America	1,675,820
Public Citizen Foundation	8,482,766
Public Citizen, Inc.	4,478,746
Rainforest Action Network	2,132,810
Sierra Club Foundation	42,656,970
Social & Environmental Entrepreneurs	3,539,608
Tides Center	62,540,112
Turning Point Project (2000)	634,664
Union of Concerned Scientists (2003)	8,753,320
US Public Interest Research Group	4,497,633
Water Keepers Alliance (2003)	1,251,263
Total annual expenditures of leading antibiotechnology activist groups	**627,833,586**

SOURCE: Publicly filed IRS 990 forms, available at www.guidestar.org.

of financial connections uncovers an extensively linked network. The top antibiotechnology activists (anticorporate, anti–free trade, antiglobalization, and so forth) have ties to a relatively small number of people directing a large number of organizations with control over a significant amount of resources.

Consider the web of influence that emanates from the Center for Food Safety (CFS), run by Andy Kimbrell, Joseph Mendelsohn, and Adele Douglas, often coordinates antibiotechnology campaigns and litigation. Douglas and Kimbrell also run the International Center for Technology Assessment (ICTA), which is located in the same Washington, D.C., office as CFS. Kimbrell and Mendelsohn also run the Turning Point Foundation (same location) with Jerry Mander. Mander is affiliated with the Foundation for Deep Ecology (which funds as much as 90 percent of CFS projects along with numerous other antibiotechnology projects), the International Forum on Globalization (a project of the Tides Center), Environmental Grantmakers Association, and a public interest communication and advertising agency providing support to antibiotechnology groups. Douglas and Kimbrell also share ties to the American Humane Association and the Humane Society of the United States. Kimbrell is also a director of International Forum on Globalization (Jerry Mander again) and is listed as a board member of the antibiotechnology Institute for Agriculture and Trade Policy (IATP), run by Mark Ritchie, and the Foundation on Economic Trends (FOET), run by antibiotechnology activist Jeremy Rifkin. The combined 2001 annual budgets for the groups that Kimbrell helps direct is nearly $18.2 million. The resources controlled by his associates exceed $100 million.

CFS has spawned other groups, including the Organic Consumers Association (OCA), run by former CFS and FOET staffer Ronnie Cummins, and Organic Watch, run by the American Humane Association director and organic industry lobbyist Roger Blobaum. OCA was an internal CFS campaign until 1998, when it was spun off. While still part of CFS, OCA's financial sponsors included Eden Organic and Patagonia Clothing. The for-profit law firm of Kimbrell & Mendelsohn is listed on public IRS tax forms as a financially linked partner to their International Center for Technology Assessment (ICTA) and CFS antibiotechnology campaign organizations. CFS's advisory board also includes representatives from Rodale's Organic Gardening, the Organic Certifiers, and the leading registered lobbyists in Washington, D.C., for the organic industry.

Numerous CFS Web site domain names are registered to West Coast OCA coordinator Steve Urow. Urow, along with OCA director Ronnie Cummins, is also listed as the registrant and a principal of GreenPeople.org, a supplier of organic and "socially responsible" products and a project of Social and Environmental Entrepreneurs. GreenPeople's other Web campaigns include MonsantoSucks.com and the SRI site, CompanyEthics.com. CFS is also listed as a founding member of the "Genetically Engineered Food Alert" campaign launched by Fenton Communications (which also runs Environmental Media Services), the most prominent activist public relations agency, based in Washington, D.C. Fenton's client list includes numerous other antibiotechnology activists such as Greenpeace, several organic and natural product companies, socially responsible investment funds, and the grant-making foundations that fund the efforts. This dense web of activists is illustrated in figure 7-1.

As the CFS example demonstrates, one group and a handful of individuals are tied to dozens of other so-called independent organizations, often promoted as representing diverse opposition to biotechnology. These individuals control tens of millions of dollars (in some cases hundreds of millions) in advocacy resources and coordinate the activities of dozens of front organizations. The vast majority of formally incorporated antibiotechnology groups share common board members, staffers, and/or funding sources. A larger number of informal antibiotechnology organizations are formed and dissolved every year by these same people, allowing the groups to avoid prolonged public scrutiny and having to file tax and other financial records publicly.

In addition to employing activists, traveling the globe to attend protests, and building networks of websites, this loosely interlocked network spends millions of dollars on well-paid lobbyists, professional public relations firms (like Fenton), and a range of outreach activities, including multi–million-dollar advertising campaigns, telemarketing, and sophisticated training programs on issues ranging from creative confrontation with police to Internet propaganda. News reports suggesting that opposition to biotechnology is the result of a grassroots consumer movement and claims that activists are protesting on a dollar a day ring ever more hollow.

Andy Kimbrell's network and influence at CFS are not unique. His associates and business partners have similar, interconnected, and often

Figure 7-1
Organizational and Funding Links among Key Activist Group Leaders

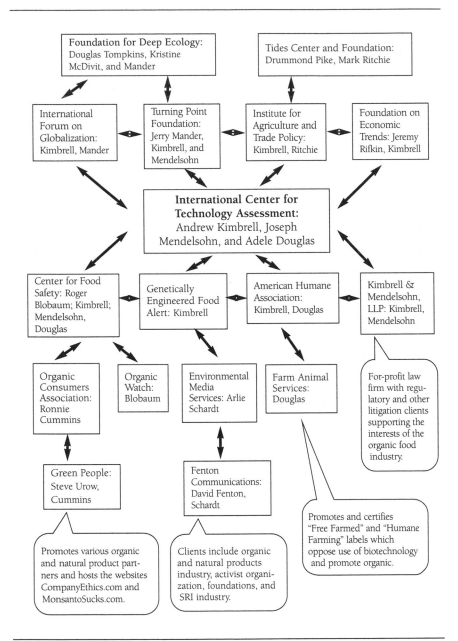

undisclosed potential conflicts of interest with the various campaigns they help direct. Combined, these tax-exempt groups and individuals represent a multi-billion-dollar protest industry promulgating a vast array of for-profit side businesses, all of which have direct links and financial ties to the organic and natural products industry.

Marketing

The organic and natural products movement and the movement for corporate social responsibility and socially responsible investment use anti-biotechnology rhetoric and support for activist groups to validate their products and increase their markets. The combination of advertising, public relations campaigns, lobbying, and financial support for synergistic protest groups has created an extremely negative environment for agricultural biotechnology that was unanticipated by its supporters.

The Organic and Natural Products Industry. The organic and natural products industry has the combined motivation of protecting existing markets from less-expensive foods produced by biotechnology (often competing with organic as "pesticide-free"), and increasing organics' existing market share (Forrer, Avery, and Carlisle 2000). Consider the "cause-marketing" campaign launched by Patagonia Clothing, whose advertising emphasizes its "social responsibility" and use of organic cotton while the Patagonia Foun-dation funds antibiotechnology efforts. Patagonia advertising and in-store displays, for example, included highly critical attacks on biotech corn and allegations that it was dangerous to butterflies. The founder of Patagonia Clothing, Kristine McDivitt, also heads the nonprofit Foundation for Deep Ecology, which funds groups like the Turning Point Foundation that conveniently target Patagonia's GM-using competitors.

Key analysts have pointed out the synergy that exists between anti-biotechnology activists and "natural" product companies. Thomas Hoban of North Carolina State University calls opposition to biotechnology "a key marketing strategy for the organic industry" (Hoban 2000). He quotes David Martosko, director of research for the Center for Consumer Freedom, who says, "Support for food scaremongers comes from organic and natural food marketers, eager to hurt their conventional competitors and build market share." That's not to say, of course, that McDivitt and other business activists do not hold genuine, strong opinions in opposition to biotechnology. However, this synergy does highlight the conflict of interests that permeates the "social responsibility" community and the organics industry.

The two largest retailers of natural and organic foods, Wild Oats Markets and Whole Foods Markets (which promotes itself on its Web site home page as believing in "a virtuous circle entwining the food chain, human beings and Mother Earth . . . through a beautiful and delicate symbiosis"), closely align themselves with antibiotechnology activists, often sponsoring protest events such as the annual "Biodevastation" conferences, and initiatives like the "Campaign to Label Genetically Modified Foods."

The organic community is well aware of the financial benefits that come with discrediting less-expensive conventional and biotech products. At the 1999 Organic Food Conference, growers and retailers were warned by a leading industry marketing expert that "if the threats posed by cheaper conventionally-produced products are removed, then the potential to develop organic foods will be limited." That same year, the organizer of the annual Summit on Organic Food Technology told participants, "Right now, Europe is freaking out about genetically altered produce. That's an opening for U.S. organic growers" (Forrer, Avery, and Carlisle 2000). In an analysis entitled, "Why Organic Has the Power to Alter All Food Marketing," the journal *Marketing* noted, "Years of black propaganda from the organic fringe, backed by the mad cow tragedy, have had a perceptible effect on consumer views" (*Marketing* 1999).

Trade groups, such as the Institute for Agriculture and Trade Policy (IATP), receive funding and support from a wide range of organic groups, outwardly mainstream philanthropic interests, and corporations. IATP's Mark Ritchie and its leadership also consult with and provide counsel to such leading biotechnology commercial interests as Dupont. Yet, in practice, IATP has morphed into a leading antibiotechnology advocacy group.

Like many antibiotech activists, IATP publicly professes a desire to work with corporations to make them better global citizens, but its behind-the-scenes actions suggest a confrontational ideology. For example, in May 2003, the group simultaneously helped organize and joined protests seeking to disrupt the World Agricultural Forum in St. Louis, while at the same time sitting on the Forum's NGO advisory board, ostensibly to help make the conference successful. IATP's leadership has its hands in dozens of other protest campaigns, and has even shared resources with extremist groups like Earth First!, which takes credit on its Web site for "serious sabotage and damage" against government offices; destruction and theft of equipment, records, and other property; and "stock destroyed" in nighttime raids on agricultural facilities.[1]

In addition to its numerous organic lobbying campaigns, IATP also owns Headwaters International, a trading company that deals in organic coffee. Also known as "Peace Coffee," Headwaters receives tens of thousands of dollars in payments from the nonprofit Tides Center, which also funds the IATP-supported Organic Consumers Association campaign, targeting Starbucks to buy more organic coffee from suppliers like Headwaters (Forrer, Avery, and Carlisle 2000). (IATP's President Mark Ritchie is also noted as the registered agent on Tides's incorporation filing documents in the state of Minnesota.)

CSR/SRI. The corporate social responsibility and socially responsible investment communities are replete with similar conflicts of interest, including ties to organic activists. As Mark Ritchie noted in an IATP-circulated e-mail in April 2000, "Financial institutions provide campaigners with a key strategic lever—if successful, financial campaigns have the power to withdraw necessary funding from the companies, cutting off the unwanted technology at the source" (Ritchie 2000).

Many for-profit companies that promote themselves as "socially responsible" and position themselves as icons of the SRI industry have launched cause-marketing promotions targeting biotechnology. The Body Shop, Ben & Jerry's, Tom's of Maine, Stonyfield Farms, Whole Foods Markets, and Working Assets, a credit card and telecommunications company that touts itself as "created to build a world that is more just, humane and environmentally responsible," are aggressive donors and advertisers. These companies often fund and coordinate with radical activists direct-action campaigns that have included personal assaults and criminal destruction of property. Whether intended or not, their effect is to spread unfounded public fears to undermine the safety and/or social reputation of competing products.

CSR and SRI groups have focused increasing and overwhelmingly negative attention on the biotechnology issue, often linking up with aggressive antibiotechnology activists. Some organizations, such as As You Sow,

TABLE 7-3

SOURCES OF ANTIBIOTECH ACTIVIST FORM LETTERS

Source	PR Firm Affiliation	Number of Form Letters
Working Assets	Fenton Communications	35,989
Sustain USA Organic Trade Association	Environmental Media Services (EMS) partner	20,656
Rodale's Publishing	Fenton Communications	17,338
Organic Farmers Marketing Association	EMS partner	11,667
Save Organic Standards Center for Food Safety	EMS partner	11,349
Mothers for Natural Law	EMS partner	4,594
PureFood Campaign Center for Food Safety	EMS partner	2,885
Center for Science in the Public Interest	Fenton Communications	1,977

SOURCE: U.S. Department of Agriculture (1997).
NOTE: EMS and Fenton share staff, office space, and other resources. EMS partners participate in Fenton-supported press conferences and other public relations activities. EMS is noted as a nonprofit project of the Tides Center, which provides tax-deductible, "donor-directed" funds to EMS from such entities as Working Assets and other donor groups (many of whom are also Fenton clients). Working Assets founder Drummond Pike runs Tides. Tides' 1998 IRS form 990 tax return notes Fenton Communications as the second-largest recipient of these tax-deductible funds in payment for services.

Alternative Power Words to Use: Controlling the Language," was directed to "activists" and "journalists," including Greenpeace, IATP, and the Organic Consumers Association. It outlined what has become a successful strategy to demonize biotechnology. Written by an organic- and natural-product advertising executive, the memo urged that "we should never use" such neutral, science-based terms as "biology," "food scientists," "biotechnology companies," or "biotechnology." The author provided a glossary of alternatives— highly negative phrases such as "genetic pollution," "Frankenfoods," "terminator seeds," "genetic engineering industry," "genetically engineered foods," "test-tube food, and "mutated food," which have become popular among headline writers and even mainstream news publications.

"Make them use our words," writes Peter Michael Ligotti (1999), the architect of several antibiotech Internet campaigns. "Look how successful the 'terminator' seed term was. At first, that was a term only used by activists. And congratulations on the success of the term 'Frankenstein food.' I am suggesting an extension of those two great successes."

Craig Winters, lobbyist and marketing consultant for the organic and natural products industry, runs the "Campaign to Label GE Foods," which is funded by the organic industry. He has noted the importance of using the key buzzwords preferred by activists, "genetically modified" and "genetically engineered," rather than "biotechnology." In an e-mail to organic industry supporters and activists on ways to "fine tune our marketing approaches," entitled "Genetic Terminology," Winters wrote, "The main thing to keep in mind is that the corporations . . . never refer to these terms. They always say 'biotechnology.' Market surveys show that consumers have a negative response to any term that includes the word 'genetically'" (Winters 1999).

This disciplined Orwellian campaign has worked. A comparison of articles from the first six months of 1993—the year the first biotech crops were approved for commercial use—with those during the first six months of 2002 suggests that activists have successfully defined the terms of public dialogue on biotechnology. There was a one-hundred-fold increase in the media's use of the more inflammatory and emotional words such as "genetic," "manipulation," and "altered" as compared to more neutral terms such as "biotechnology" or "bioengineered."

Table 7-4 shows the results of a search-term frequency review of key phrases associated with agricultural biotechnology appearing in the major search engines. Conducted in April 2003 using Overture and Wordtracker services, the review found negative terms outnumbering neutral ones by more than ten to one.

Search engine results are also heavily weighted with protest groups and organic marketing campaign sites uniformly critical of agricultural biotechnology. Using the most frequently searched terms—"genetically modified," "genetically engineered," or "genetically altered"—not one biotechnology corporate or industry trade association website comes up in the top fifty results. Even when using the more neutral biotechnology phrases, the results include

TABLE 7-4

FREQUENCY OF KEY WORDS IN FOOD BIOTECHNOLOGY DEBATE

Term/Phrase	Frequency
Genetically modified (food, crop, plant, etc.)	20,080
Genetically engineered (food, crop, plant, etc.)	7,508
Genetically altered (food, crop, plant, etc.)	2,827
Frankenfood(s)	461
Terminator (seed, gene, plant)	205
Total searches per month using negative terminology	**31,081**
(Agriculture, plant, food, etc.) biotech(nology)	2,807
Total searches per month using positive terminology	**2,807**

SOURCE: Overture and Wordtracker services, April 2003.

a significantly higher proportion of critical destinations in the top results. The Internet-consuming public is significantly more likely to be influenced by opponents of biotechnology than either supporters or neutral stakeholders.

Steven Katz, an associate professor of English at North Carolina State University, studies the linkage between language and public perceptions of risk. "The important role that language plays in the public's perception and reception of scientific data and risk assessment is often neglected by scientists," says Katz. He singles out issues that have been slowed or completely halted by public concerns driven by language—including biotechnology. He notes that such "public resistance has been traced to communication problems—flawed rhetorical choices and faulty assumptions by scientists about the role of language, emotion and values in communicating with the media and public" (Katz 2001). A study by the London School of Economics found that the media's use of "Frankenfood" headlines and other negative metaphors for foods produced using agricultural biotechnology have helped create and fuel public fears (Reuters News Service 1999). Numerous published polls show a majority of consumers support foods derived from "biotechnology," yet others show these same consumers oppose "genetic engineering" of food (*Farmers Guardian* 2001).

To illustrate how activist groups, along with what amounts to their marketing partners like the organic and CSR/SRI industries, leverage the

Internet to their advantage, we applied a proprietary data-analysis system that combines an interactive graphical interface with a comprehensive database of online resources linked to a particular issue or product. Using a series of algorithms and other weighting factors, we generated a picture of the online environment for any product, issue, or demographic group.

Online environment analysis conducted in June 2002 by v-Fluence Interactive Public Relations reveals the online environment for one well-known biotechnology-related consumer food product to be heavily dominated by negative influencers. Fifty-seven percent of the overall online environment for this product was negative and much better positioned to be found by the Internet-using public than supportive or neutral destinations. The negative stakeholders shared similar content and frequently linked to one another. This was in stark contrast to supportive destinations, which did not share content and had few links to and among one another. To view the graphical representations of these i-Maps™, visit http://www.v-fluence.com/news/aei.html.

These i-Map images illustrate how activists can dominate the online environment through simple, but extremely effective, tactics. The Internet has become the primary source of information on a range of topics. At the same time, numerous surveys confirm its influence over a majority of key stakeholders, including journalists, government regulators, and other opinion leaders, and note that the Internet's power now exceeds that of any other medium.[3] Activists' dominance of the online environment enables them to exert outsized influence over the broad public dialogue associated with various protest topics, agricultural biotechnology being a case in point.

Charting a Course for Change

Money plus marketing plus the Internet add up to significant influence for the antibiotechnology industry. Yet it is still perceived by the public as a grass-roots movement. These factors remain significantly underreported by the media and largely unappreciated by the biotechnology industry. Biotechnology supporters and commercial interests need to change significantly the way they have responded to these groups.

As a critical first step, they need to learn from the activists and use the Internet effectively. Independent research and statements in support of biotechnology should be freed from copyright restrictions so they can be replicated in multiple destinations and linked to and from corporate, trade association, and other sites. Competitors must also work together to reinforce each other's successes directly; relying solely on trade associations to defend common industry interests is not sufficient and often results in defining the entire industry by its weakest links. Pressure needs to be applied on Better Business Bureaus, professional associations, and state and federal regulators to hold organic and natural product companies accountable when their advertising claims are blatantly misleading or false. The multi-billion-dollar organic and CSR/SRI industries should be challenged to meet the letter of the law regarding their masking of marketing expenses as tax-deductible contributions.

If perception is reality, then the antibiotechnology movement has won many battles. The war for the public hearts and minds is far from over regarding agricultural biotechnology. Perceptions can be changed. Indeed, they must be changed, if the promises of biotechnology are to become reality.

Notes

1. Agbioworld.org reports that in 2001 IATP's Sustain and "Project GEAN" (Genetic Engineering Action Network) campaigns shared the same Chicago mailing address as Earth First! http://www.agbioworld.org/newsletter_wm/index.php?caseid=archive&newsid=990.

2. ICCR proposes resolutions primarily targeting two industries: life science companies, to stop commercializing genetically engineered foods (GEF) until long-term safety is assured; and food companies, restaurants, and supermarkets, to remove genetically engineered (GE) ingredients from foods sold under house-brand or private labels until long-term safety testing is assured, with the interim step of labeling products that contain those ingredients. CSRwire is a project of SRI World Group, which also runs the Shareholder Action Network (SAN, a project of the Social Investment Forum, the SRI industry trade group), which has directed campaigns at Monsanto for its development of agricultural biotechnology products. SAN has also sponsored or supported dozens of antibiotech shareholder resolutions in recent years, many of which target major corporations such as Monsanto, Sara Lee, and Kraft.

3. Various studies found on www.nua.ie and www.cyberatlas.com, e.g., http://www.clickz.com/stats/sectors/advertising/article.php/1460711.

References

Efund. 2000. Efund Report. http://www.efund.com/Live/pdfs/IFSP2000.pdf (accessed June 2002).

Farmers Guardian. 2001. Majority Oppose GM Foods (70% Oppose). December 28.

Forrer, Grady, Alex Avery, and John Carlisle. 2000. Marketing & the Organic Food Industry. http://www.stoplabelinglies.com/pdf/organicmarketing.pdf.

GE Euphemisms and More-Accurate Alternative Power Words to Use: Controlling the Language. 1999. Circulated e-mail memo. http://home.intekom.com/tm_info/rw90525.htm.

Green Century Funds. 2002. Environmental Investing Strategy. http://www.greencentury.com/ER.htm.

Hoban, Thomas J. 2000. Countering the Myths of Biotechnology Acceptance. *Grand Forks Herald.* August 8.

IFIC Backgrounder. 2002. Support for Food Biotechnology Holds (71% Support). September 23. www.ific.org.

Katz, Stephen. 2001. Language and Persuasion: The Communication of Biotechnology with the Public. *AgBioForum* 4:93–97.

Ligotti, Peter. 1999. For Activists/Journalists—GE Power Words to Use: Controlling the Language. Ban GE e-mail listserv. May 17.

Manheim, Jarol B. 2001. *The Death of a Thousand Cuts.* Mahwah, N.J.: LEA Publishers, xiv.

Marketing. 1999. Why Organic Has the Power to Alter All Food Marketing. April 29.

Martosko, David. 2003. Food Fetish. *The American Spectator.* January/February.

Nunberg, Geoffrey. 2003. As Google Goes, So Goes the Nation. *New York Times.* Week in Review. May 18.

Reuters News Service. 1999. Frankenfood Headlines Scare Public, Study Shows. July 16.

Ritchie, Mark. 2000. IATP E-mail Listserv Message. April.

U.S. Department of Agriculture. 1997. Web site report on public comments received. http://www.ams.usda.gov/nop/archive/1997Comments/1997FormLetters.html (accessed April 22, 2005).

v-Fluence Interactive Public Relations. 2002. Online Environment Analysis and i-Maps. June. http://www.v-fluence.com/news/aei.html.

White, Clem. 1999. Environmental Activism and the Internet. Master's thesis. Massey University (New Zealand). December. http://www.arachna.co.nz/thesis/home.asp (accessed June 2002).

Winters, Craig. 1999. Genetic Terminology. BAN-GEF listserv. May 26. www.beyond-the-illusion.com/files/New-Files/990630/geneticTerminology.txt.

Working Assets. 2002. "About us" Web page. http://www.workingassets.com/aboutwa.cfm (accessed June 2002).

8

"Functional Foods" and Biopharmaceuticals:
The Next Generation of the GM Revolution

Martina Newell-McGloughlin

Biotechnology has introduced a new dimension of innovation to agricultural productivity. It offers efficient, reliable, and cost-effective means to produce a diverse array of products and tools with the potential to improve food, feed, and fiber production, reduce the dependency of agriculture on chemicals, and lower the cost of raw materials, all in an environmentally sustainable manner.

Despite some vocal assertions to the contrary, commercialization of the first generation of products of recombinant DNA technology is just another chapter in a long history of human intervention in nature for such purposes. Such innovation must be undertaken within a regulatory framework that appropriately manages risk and ensures adequate protection of the consumer and the environment, while not stymieing the development of new products and processes.

Propagation of the next generation of plants will focus on identifying and isolating genes and metabolites that will make possible the enhancement of valuable traits, with some of the later compounds being produced in mass quantities for niche markets. Two of the more promising markets are nutraceuticals, or so-called functional foods, and plants developed as bioreactors (production factories) for the production of valuable proteins and compounds, a field known as plant molecular farming.

Functional Foods

A functional food is defined as any modified food or food ingredient that may provide a health benefit beyond the traditional nutrients it contains (Institute of Medicine 1994, 109). The term *nutraceutical*, coined in 1989 by Stephen De Felice, founder and chairman of the Foundation for Innovation in Medicine, refers to "any substance that may be considered a food or part of a food and provides medical or health benefits, including the prevention and treatment of disease" (Brower 1998).

Epidemiological research has found increasing evidence that eating functional foods leads to better health. They have been associated with the prevention and treatment of at least four of the leading causes of death in the United States: cancer, diabetes, cardiovascular disease, and hypertension. The National Cancer Institute estimates that one in three cancer deaths is diet related, and that eight of ten cancers have a nutrition or diet component (Steinmetz and Potter 1996).

Plants are remarkable in their capacity to synthesize a variety of organic substances, such as vitamins, sugars, starches, and amino acids. As many as 80,000 to 100,000 different substances are synthesized in plants, including macronutrients (e.g., proteins, carbohydrates, lipids [oils], and fiber), micronutrients (e.g., vitamins and minerals), antinutrients (e.g., compounds such as phytate that reduce bioavailability allergens [e.g., albumin]), endogenous toxicants (e.g., glycoalkoloids and cyanogenic glycosides), and other plant-specific compounds (some of which may have beneficial effects) that are significant to human and animal health. Both traditional plant-breeding and biotechnology techniques are needed to overcome a variety of technical challenges inherent to metabolic engineering programs and produce plants carrying traits that enhance them as functional foods. Intricate functional modifications will require more sophisticated plant manipulation than the simple (relatively speaking) addition or removal of one or two genes as is the case for many agronomic traits. In many instances, complex trait modification will require engineering of metabolic pathways (i.e., series of biochemical reactions inside of a living organism that result in the transformation of a particular molecule into a different molecule), which is a much more challenging prospect, as one is dealing with complicated interconnected networks within the organism.

Continuing improvements in molecular and genomic technologies are speeding up the development process. Significant progress has been made in recent years in the molecular dissection of many plant pathways and in the use of cloned genes to engineer plant metabolism. Regulatory oversight of engineered products has been designed to detect any unexpected outcomes in biotech crops and, as more metabolic modifications are made, new methods of analysis are being developed to address these issues.

A number of advances have been made in recent years toward enhancing the healthful qualities of specific varieties of foods and food components. Biotechnology has provided powerful tools, for example, for modifying the composition of oilseeds to improve their nutritional value and provide the functional properties required for various food oil applications, such as greater stability during frying and longer shelf life to prevent rancidity. Edible oils rich in monounsaturated fatty acids provide improved oil stability, flavor, and nutrition for human and animal consumption. Turning off the expression of the enzymes oleate desaturase in soybean resulted in oil that contained greater than 80 percent oleic acid (23 percent is normal) and had a significant decrease in polyunsaturated fatty acids (Kinney and Knowlton 1998). High-oleic soybean oil is naturally more resistant to degradation by heat and oxidation, and so requires little or no postrefining processing (hydrogenation), depending on the intended vegetable oil application.

In 1999, using nutritional genomics, Della Penna isolated a gene by means of which the vitamin E content of Arabidopsis seed oil was increased nearly tenfold. Progress has been made in applying this technology to crops such as soybean, maize, and canola, as well as extending it to increase folates in rice. It also has shown potential for producing industrial oils and chemicals in genetically engineered crops.

Another notable success has been the engineering by Ursin (2000) of canola to produce omega three fatty acids (James et al. 2003), which could help in the regulation of inflammatory immune reactions and blood pressure, treatment of conditions such as cardiovascular disease and cystic fibrosis, brain development in utero, and the development of cognitive function in infancy. This advance supplements the currently limited supply of fish oil, previously the critical source of these fatty acids.

Humans, as well as poultry, swine, and other nonruminant animals, have specific dietary requirements for each of the essential amino acids. A deficiency of one essential amino acid limits growth and can be fatal. This is a serious consideration for vegetarians whose whole source for proteins is plant-based. In animal feeds, the primary limitations of maize and soybean meal-based diets are for lysine in nonruminant mammals and methionine in avian species. Through modern biotechnology, maize with increased levels of lysine and soybeans with increased levels of methionine could allow diet formulations with improved amino acid balance, without the need to add crystalline lysine and methionine.

Similarly, a team led by Ingo Potrykus (1999) engineered so-called Golden Rice to produce pro–vitamin A, which is an essential micronutrient. Widespread dietary deficiency of this vitamin in Asian countries where rice is the staple food predisposes children to diseases such as measles, with tragic consequences. Improved vitamin A nutrition would alleviate serious health problems and, according to UNICEF, could also prevent up to 2 million infant deaths from vitamin A deficiency (UNICEF 1997, 2000).

Despite all these promising developments in food biotechnology, scientific research is still needed to confirm the benefits of any particular food or component. For functional foods to deliver their potential public health benefits, consumers must have a clear understanding of, and a high degree of confidence in, the scientific criteria that are used to document health effects and claims. Because decisions about this technology will require an understanding of plant biochemistry, mammalian physiology, and food chemistry, strong interdisciplinary collaborations will be needed among plant scientists, nutritionists, and food scientists to ensure a safe and healthful food supply.

Plant Molecular Farming

In addition to being a source of nutrition, plants have been a valuable wellspring of therapeutic substances for centuries. During the past decade, intensive research has focused on expanding this source through recombinant DNA biotechnology and, essentially, using plants and animals as living factories for the commercial production of vaccines, therapeutics, and other valuable products such as industrial enzymes and biosynthetic feedstocks, as shown in table 8-1.

TABLE 8-1
POSSIBLE THERAPEUTICS MADE FROM MOLECULAR FARMING

1) Primary Products	2) Derived Products
Monoclonal antibodies (Mabs) (Passive immunity): Herceptin for breast cancer	Bio-plastics: polyhydroxybutyrate, a thermostable, biodegradable plastic
Vaccines (Active immunity): Edible vaccines for hepatitis B and traveller's diarrhea	Nutraceuticals:
Structural: proteins, peptides, matrices: collagen, gelatin	Macro: Carbohydrates, fats, fiber Micro: Vitamins, cofactors, minerals
Enzymes: food, feed, industrial, therapeutic, diagnostic, cosmetic, trypsin for many industrial processes, including the manufacturing of human health products	Phytochemicals: Carotenoids, flavonoids, isoflavones, isothio-cyanates, phenolics, tannins
Antidisease therapeutics: Factor VII for hemophilia aprotinin to limit blood loss	Fragrances, flavors
Enzyme inhibitors	Fibers: polymers, lignins, structural starches, Polylactide for fleece, Cellulosic Nano Fibers

SOURCE: Compiled by author.

Possibilities in the medical field include a wide variety of compounds, ranging from edible vaccine antigens against hepatitis B and Norwalk viruses (Arntzen 1997; Dixon and Arntzen 1997; Mason et al. 1996, 1998; Richter et al. 2000) and infectious bacteria (Brennan et al. 1999) to vaccines against cancer and diabetes (McCormick et al. 1999); enzymes (Hogue et al. 1990; Verwoerd et al. 1995); hormones (Leite et al. 2000; Staub et al. 2000); plasma proteins (Sijmons et al. 1990); and human alpha-1-antitrypsin (Terashima 1999). The company FibroGen is looking at plant-based systems to inexpensively mass produce alternatives to animal-based sources of collagen and gelatin which have extensive applications in everything from gel capsules to burn treatment to cosmetic enhancement.

Plant cells are capable of producing a large variety of recombinant proteins and protein complexes. Therapeutics produced in this way are termed "plant-made pharmaceuticals" (PMPs); nontherapeutics are termed plant-made industrial products (PMIPs). This discussion will primarily focus on PMPs, but many aspects can, of course, equally apply to PMIPs.

Two of the seminal papers that gave proof of the concept of the use of transgenic plants as pharmaceutical production systems appeared in 1995 (Haq et al. 1995; J. K.-C. Ma et al. 1995, 1998; S.-W. Ma et al. 1997). In 1998, the same two groups reported the results of successful human clinical trials with their transgenic plant-derived pharmaceuticals: an edible vaccine against *E. coli*–induced diarrhea and a secretory monoclonal antibody directed against *Streptococcus mutans*, for preventative immunotherapy to reduce incidence of dental caries. Haq et al. (1995) reported the expression in potato plants of a vaccine against *E. coli* enterotoxin (ETEC) that provided an immune response against the toxin in mice. Human clinical trials suggest that oral vaccination against either of the closely related enterotoxins of *Vibrio cholerae* and *E. coli* induces production of antibodies that can neutralize the respective toxins by preventing them from binding to gut cells. Similar results were found for Norwalk virus oral vaccines in potatoes (Tacket et al. 2000). For developing countries, the intention is to deliver them in bananas or tomatoes (Arntzen 1997; Mason et al. 2002).

In Guys Hospital in London, Ma et al. (1995, 1998) showed that tobacco plants could express secretory antibodies or "plantibodies" against the cell surface adhesion protein of *S. mutans*. Used as a bactericidal mouthwash, the antibodies prevented bacterial colonization by the microorganism and development of dental caries for four months. A similar approach showed that soybean-produced antibodies protected mice against infection by genital herpes (Zeitlin et al. 1998). The different processing systems of plants and animals do not appear to affect the immune functions of the plant-derived antibodies, which is of real interest to companies such as Centocor, Abginex, and Immunex (now Amgen) who are looking to more effective, efficient, and economic ways to produce humanized monoclonals.

Many new pharmaceuticals based on recombinant proteins will receive regulatory approval from the U.S. Food and Drug Administration (FDA) in the next few years. As these therapeutics make their way through clinical trials and evaluation, the pharmaceutical industry faces a production-capacity challenge using existing steel bioreactors (similar to the large vats in a local microbrewery—there are not enough Anheuser Busch–like facilities in the world to meet exponentially increasing demand). Pharmaceutical discovery companies are exploring green plant–based production to overcome capacity limitations, enable production of complex therapeutic proteins, and fully realize the commercial potential of their biopharmaceuticals. As with any manufacturing process, production space is an issue. It takes five to seven years to build a biologics plant, compared to one to three years for a conventional pharmaceutical plant.

The use of green plants as bioreactors would allow production of recombinant proteins in large quantities and at relatively low costs. With sunlight and soil-based nutrients as inputs, plants produce large quantities of material. They offer the ability to increase production at low cost by planting more acres rather than building fermentation capacity, along with lower capital and operating costs compared to traditional production facilities and simplified extraction of the therapeutic compared to traditional technology.

Biopharming may also enable research on drugs that cannot currently be produced. For example, CropTech in Blacksburg, Virginia, is investigating a protein that seems to be a very effective anticancer agent. The problem is that this protein is difficult to produce in mammalian cell culture systems as it inhibits cell growth. This should not be a problem in plants.

The making of pharmaceuticals from plants, examples of which are illustrated in table 8-2, is also a sustainable process, because the plants and crops used as raw materials are renewable. And the system has the potential to address problems associated with provision of vaccines to people in developing countries. Products from these alternative sources do not require a so-called cold chain for refrigerated transport and storage, and the availability of oral vaccines obviates the need for needles and aseptic conditions that is often a problem in those areas.

Molecular farming may fall short of its potential if there are unanticipated purification problems or if therapeutic substances made in plants are not as effective as those made in mammalian cells. The development of pharma plants could also be hampered if cost savings are not realized, or if legal barriers or marketing issues arise related to the negative public perceptions that have surrounded the entire field of genetic engineering.

TABLE 8-2

EXAMPLES OF PLANT-MADE PHARMACEUTICALS

Alfalfa	Plasma proteins foot-and-mouth disease vaccine
Maize	Anti-HIV and anti–herpes simplex antibodies; Microbiocides for pulmonary infection; Mabs for cancer, autoimmune disease (rheumatoid arthritis); Vaccines for hepatitis B, Norwalk virus (Travelers disease); Vaccines and Mabs for animal disease prevention; Aprotinin for blood loss and heart surgery
Lemma	Human tissue plasminogen activator for peripheral arterial occlusion
Lettuce	Vaccines for hepatitis B
Moss (bryophyte)	Factor IX for Haemophilia B
Pigeonpea plant	Rinderpest (cattle virus)
Rice	Alternatives to antibiotics in poultry diets; Lactoferrin, Lysozyme for gastrointestinal health, topical infections, inflammations, B-cell lymphoma vaccine
Safflower	Therapeutics and oil-based products for oral/dermal delivery
Spinach	Protective antigen for vaccine against *Bacillus anthracis*
Soybean	Herpes simplex virus 2; Tobacco extensin signal peptide; Anti-HSV-2 (IgG)
Tobacco	Non-Hodgkins B-cell lymphoma TGF-b glucocerebrosidase for Gauchers Syndrome, *Alpha galactosidase* for enzyme replacement therapy, IgGs for prevention of dental decay, common cold, GAD 7 cytokines for type I diabetes, Colon cancer surface antigen, fabrazyme fat-storage disorder called Fabry disease
Tobacco and Corn	Gastric lipase for cystic fibrosis, Lactoferrin for gastrointestinal infections
Tomato, Potato, Tuber, and Banana (someday!)	Edible vaccines against Enterotoxigenic *E. coli*, Norwalk virus, Hepatitis B, *Vibrio cholera*, Rabies virus-intact glycoprotein antimicrobial peptides
Wheat	Carcinoembryonic antigen; Murine IgG signal peptide

SOURCE: Compiled by author.

There are, indeed, novel challenges to oversight, risk assessment, and public perception, the most pressing of which is to ensure adequate segregation from conventional foods to prevent unintentional commingling. The new products will be controlled as medicines; however, unlike contained bioreactors, there is potential for environmental and food safety mishaps. Especially in the case of active pharmaceutical ingredients (API), there is genuine concern of gene escape and failure to segregate from the food supply.

No matter how comprehensive the containment facilities and measures in place or how vigilant the oversight and monitoring procedures, there could be unexpected events. Some occurred after field tests in Iowa and Nebraska by ProdiGene, which produces animal vaccines in maize. In the Nebraska case, officials realized that some 500,000 bushels of harvested soybeans were contaminated ("adulterated," in the parlance of the Food, Drug, and Cosmetics Act) with small amounts of genetically modified (GM) maize, which had been grown during 2001 on the same plot, because the farmer did not weed "volunteer" tasseled plants from the field in which the soy was grown. In Iowa, federal officials required a local producer to destroy some 155 acres of maize because it could have been cross-pollinated by ProdiGene's engineered maize being raised in a nearby field. Without admitting to those violations, ProdiGene agreed to post a $1 million bond and also to reimburse the U.S. Department of Agriculture (USDA) for the costs, which ran to several million dollars, involved in disposing of the contaminated crops. These incidents underscore that, while the technology permits an attractive production, storage, and delivery system, the risks of growing such pharmaceutical maize in the midst of a region where the majority of farmland is used to produce commodity food and feed crops may outweigh the benefits.

Currently, test plots for crops engineered to grow pharmaceuticals are regulated by the USDA's permit system. USDA regulations require developers to have clearly written procedures, to handle wastes properly, and to maintain production and control records. Plant-made pharmaceuticals also come under the purview of the U.S. Food and Drug Administration. The FDA will monitor the manufacturing process as well as the purity and consistency of the products under its "good manufacturing practices" guidelines. Nevertheless, both the USDA and FDA recognize that this is a new application of the technology and are in the process of coming up with guidance specifically for pharmaceutical plants.

The growing and harvesting of the crop itself must comply with principles of confinement, which essentially means keeping the crop and its recombinant products on the land where it was grown until they are removed for processing, with no inadvertent exposure to the public and minimal exposure to workers and the environment. Misadventures such as the commingling with food maize of the nonfood, Bt-maize hybrid known as StarLink, which was only registered for animal feed, will not be repeated if these procedures and rules are followed. Precaution demands that standard operating practices be implemented for a functional identity preservation system, which ensures that the pharm crop is completely segregated from all other crops and that protocols are in place for its production and handling. Achieving this goal is possible with implementation of chain-of-custody procedures that track the product through every stage of production and processing. Remedial plans must be in place that will trigger procedures to mitigate any potential effects if the confinement system does not work.

Candidate plants for the production of PMPs include familiar crops like alfalfa, canola, maize, potato, rice, safflower, soybean, and tobacco. These food and nonfood crops are treated altogether differently from the biotechnology-derived crops designated for food purposes. Issues of safety and efficacy are evaluated when considering plants as candidates. Pharmaceutical-producing plants have been studied with respect to pollination, genetics, seed dormancy, and weediness potential. This information can be useful for addressing several concerns, including pollen movement and subsequent gene flow between conventionally bred and biotechnology-derived crops. A long history of cultivation shows that the candidate crops are the least likely to be invasive of "natural" ecosystems.

The infrastructure of regulation has been in place for nearly a decade, and it continues to evolve as experience with biotechnology-derived food crops grows. Risk management must necessarily focus on providing health protection both for humans (workers and consumers) and ecosystems. The principal concerns are for gene flow and commingling. Adulteration of the food supply with therapeutic genes will always be at a zero-tolerance level. Gene flow from PMP farms is likely to be the most contentious issue during the crop-production phase of PMP technology. Some strategies to reduce the risk of gene flow from transgenic crops, such as the use of male sterile plants, work well but are limited to a few species. For the many crops in which chloroplasts (organelles where photosynthesis occurs and which have their own DNA) are not transmitted through pollen, transformation of the chloroplast genome should provide an effective way to contain foreign genes, since if they cannot get into the pollen they cannot get out of the mother plant. But it is not foolproof, as demonstrated by papers in 2003 indicating some illegitimate recombination with genomic DNA, whereby genes inserted into the chloroplast can be transferred into the plant genome (Huang, Ayliffe, and Timmis 2003).

In the past few years, patents have been issued for techniques linking "suicide" genes to DNA "switches" that can be tripped inside pollen cells, impeding their development, and for techniques based on genes that kill off hybrid seeds as they attempt to germinate. These genetic use restriction technologies, or GURTs (pejoratively termed "terminator technology"), of which there are several, basically render crop fertility into a trait that can be switched on or off with biochemical messengers. Basically, implementation of one or more of these biological control methods can protect the food and feed supply from adventitious contamination with PMPs.

Tests to date show even when cross-pollination in maize has occurred, the probability of finding one GM seed in thousands of conventional seeds is low. It is reasonable to conclude that with proper confinement measures, any inadvertent exposures attributable to gene flow in future trials will be very low, and the risk of adverse effects to humans correspondingly low if anyone is to be inadvertently exposed through his food to an API. However, in this business, perception is reality, and the discovery, or indeed perpetration, of adulteration of the food supply despite low or no true risks could scupper the whole promising industry.

What about the persistence of therapeutic proteins in the environment? Once in the moist soil, virtually all proteins, including the pharmaceuticals, will break down into their basic constituents, amino acids. In most instances, one highly defined product will be produced in very limited acreage. For example, about one thousand acres may be required to produce enough recombinant immunoglobulins of any kind. Under optimum conditions, according to Felsot (2002), that acreage would not be concentrated in one area. Not only will a limitation in the number of acres in any one location make the land more manageable, but the specific location itself will be comparatively devoid of the food crops subject to cross-pollination.

As to the protein or protein product itself, those for use as therapeutics will already have undergone rigorous review and, in most instances, extensive clinical trials under the comprehensive safety and efficacy review and trial process given by the FDA to all pharmaceuticals, whether synthetic or biological. While other specialty crop production practices, such as for foundation seed, already have a well-tested and verified oversight, for molecular farming crops, especially PMPs, standard operating practices will be as thorough as that for stainless steel bioreactors. Oversight will be unprecedented for either agriculture or pharmaceutical production. In addition, confinement systems will be optimized to ensure adherence to the overriding requirements for identity preservation-closed loop principles. No doubt there will be mishaps, as occurred with the ProdiGene trial, but adequate remediation systems can be in place to mitigate any potential effects.

Numerous companies such as Dow and Monsanto (which have since suspended research in this area), and partnerships including smaller biotech companies, such as Centocor and Epicyte (the latter is no longer trading), have been granted permits for field trials based on reviews of detailed, credible, and practical plans for producing plant-based pharmaceuticals. In addition, only a few select growers have been identified, trained, and supervised to grow these crops, and the seeds will be available only from the manufacturer. As part of this commitment, and corporate policy, these companies will apply a comprehensive risk-focused process at key stages throughout the life cycle of every product, from conceptualization and development through to production and postproduction marketing.

While these systems hold enormous promise for all stakeholders, from the company to the producer to the patient, analytical and deliberative processes, along with proactive continuous dialogue, will be necessary among all the relevant stakeholders and partners including companies, the government, agrobiotech experts, the food industry, public policy groups, and the health care system to ensure optimum benefits for all with minimum impact on the environment and none on the food supply.

References

Arntzen, C. J. 1997. High-Tech Herbal Medicine: Plant-Based Vaccines. *Nature Biotech* 15:221–22.

Biotechnology Industry Organization (BIO). 2002. Reference Document for Confinement and Development of Plant-Made Pharmaceuticals in the United States. http://www.BIO.ORG/.

Blanas, E., et al. 1996. Induction of Autoimmune Diabetes by Oral Administration of Autoantigen. *Science* 274:1707–9.

Brennan, F. R., et al. 1999. Chimeric Plant Virus Particles Administered Nasally or Orally Induce Systemic and Mucosal Immune Responses in Mice. *Journal of Virology* 73:930–35.

Brower, B. 1998. Nutraceuticals: Poised for a Healthy Slice of the Market. *Nature Biotechnology* 16:728–33.

Cabanes-Macheteau, M., et al. 1999. N-Glycosylation of a Mouse IgG Expressed in Transgenic Tobacco Plants. *Glycobiology* 9:365–72.

Canadian Food Inspection Service. 2001. Plant Molecular Farming—Discussion Document. http://www.inspection.gc.ca/english/plaveg/pbo/mainplnt_ mfe.shtml.

Carpenter, J., et al. 2002. Comparative Environmental Impacts of Biotechnology-Derived and Traditional Soybean, Corn, and Cotton Crops. Ames, Iowa: Council for Agricultural Science and Technology. http://www.cast-science.org.

Carrillo, C., et al. 2001. Induction of a Virus-Specific Antibody Response to Foot and Mouth Disease Virus Using the Structural Protein VP1 Expressed in Transgenic Potato Plants. *Viral Immunology* 14:49–57.

Chen, Z., et al. 2003. Increasing Vitamin C Content of Plants through Enhanced Ascorbate Recycling. *PNAS* 100:3525–30.

Daniell, H., S. J. Streatfield, and K. Wycoff. 2001. Medical Molecular Farming: Production of Antibodies, Biopharmaceuticals and Edible Vaccines in Plants. *Trends in Plant Science* 6:219–26.

Della Penna, D. 1999. Nutritional Genomics: Manipulating Plant Micronutrients to Improve Human Health. *Science* 285:375–79.

Dixon, R. A., and C. J. Arntzen. 1997. Transgenic Plant Technology Is Entering the Era of Metabolic Engineering. *Trends in Biotechnology* 15:441–44.

Dry, Lisa. 2002. The Case for Plant-Made Pharmaceuticals. *BIO News*. April/May.

Ellstrand, Norman. 2002. Mindful Management of Genes That Produce Industrial Biochemicals in Plants. In *Pharming the Field: A Look at the Benefits and Risks of Bioengineering Plants to Produce Pharmaceuticals*. Proceedings from a Workshop Sponsored by the Pew Initiative on Food and Biotechnology, the U.S. Food and Drug Administration, and the Cooperative State Research, Education and Extension Service of the U.S. Department of Agriculture. http://pewag biotech.org/events/0717/ ConferenceReport.pdf (accessed January 15, 2005).

Falco, S., et al. 1995. Transgenic Canola and Soybean Seeds with Increased Lysine. *Bio/Technology* 13:577–82.

Felsot, Allan. 2002. Pharm Farming: It's Not Your Father's Agriculture. *AENews*, no. 195. http://aenews.wsu.edu.

Gerhauser, C. 1997. Cancer Chemopreventive Potential of Sulforamate, a Novel Analogue of Sulforaphane That Induces Phase 2 Drug-Metabolizing Enzymes. *Cancer Research* 57 (January 15): 272.

Goldberg, I., ed. 1994. *Functional Foods, Designer Foods, Pharmafoods, Nutraceuticals.* New York: Chapman & Hall.

Halsey, M. E., et al. 2002. Pollen-Mediated Gene Flow in Maize as Influenced by Time and Distance. Abstract and poster presented at the Annual Meeting of the American Society of Agronomy. November.

Haq, T., et al. 1995. Oral Immunization with a Recombinant Bacterial Antigen Produced in Transgenic Plants. *Science* 268:714–16.

Hipskind, J. D., and N. L. Paiva. 2000. Constitutive Accumulation of a Resveratrol-Glucoside in Transgenic Alfalfa Increases Resistance to Phoma Medicaginis. *Molecular Plant-Microbe Interactions* 13:551–62.

Hogue, R. S., et al. 1990. Production of a Foreign Protein Product with Genetically Modified Plant Cells. *Enzyme Microbe Technology* 12:533–38.

Howard, J. 1999. Plant-Based Production of Xenogenic Proteins. *Current Opinion in Biotechnology* 10:382–86

Huang C. Y., M. A. Ayliffe, and J. N. Timmis. 2003. Direct Measurement of the Transfer Rate of Chloroplast DNA into the Nucleus. *Nature* 422 (6927):72–76.

Institute of Medicine. 1994. *Opportunities in the Nutrition and Food Sciences*, ed. P. R. Thomas and R. Earl. Washington, D.C.: National Academy Press.

James, M. J., et al. 2003. Metabolism of Stearidonic Acid in Human Subjects: Comparison with the Metabolism of Other n-3 Fatty Acids. *The American Journal of Clinical Nutrition* 77:1140–45.

Jung, W., et al. 2000. Identification and Expression of Isoflavone Synthase, the Key Enzyme for Biosynthesis of Isoflavones in Legumes. *Nature Biotechnology* 18:208–12.

Kinney, A. J., and S. Knowlton. 1998. Designer Oils: The High Oleic Acid Soybean. In *Genetic Modification in the Food Industry*, ed. S. Rollerand and S. Harlander. 193–213. London: Blackie.

Lai, Jinsheng, and Joachim Messing. 2002. Increasing Maize Seed Methionine by mRNA Stability. *The Plant Journal* 30:395–402.

Leite, A., et al. 2000. Expression of Correctly Processed Human Growth Hormone in Seeds of Transgenic Tobacco Plants. *Molecular Breeding* 6:47–53.

Ma, J. K.-C., et al. 1995. Generation and Assembly of Secretory Antibodies in Plants. *Science* 268:716–19.

———. 1998. Characterization of a Recombinant Plant Monoclonal Secretory Antibody and Preventative Immunotherapy in Humans. *Nature Medicine* 4:601–6.

Ma, S.-W., et al. 1997. Transgenic Plants Expressing Autoantigens Fed to Mice to Induce Oral Immune Tolerance. *Nature Medicine* 3:793–96.

Mason, H. S., et al. 1996. Expression of Norwalk Virus Capsid Protein in Transgenic Tobacco and Potato and Its Oral Immunogenicity in Mice. *Proceedings of the National Academy of Science USA* 93:5335–40.

———. 1998. Immunogenicity of a Recombinant Bacterial Antigen Delivered in a Transgenic Potato. *Nature Medicine* 4:607–9.

Mason, H. S., H. Warzecha, T. Mor, and C. J. Arntzen. 2002. Edible Plant Vaccines: Applications for Prophylactic and Therapeutic Molecular Medicine. *Trends in Molecular Medicine* 8:324–29.

McCormick, A. A., et al. 1999. Rapid Production of Specific Vaccines for Lymphoma by Expression of the Tumor-Derived Single-Chain fv Epitopes in Tobacco Plants. *Proceedings of the National Academy of Science USA* 96 (January 19): 703–8.

Modelska, A., et al. 1998. Immunization against Rabies with Plant-Derived Antigen. *Proceedings of the National Academy of Science USA* 95:2481–85.

Mor, T. S., et al. 2003. Geminivirus Vectors for High Level Expression of Foreign Proteins in Plant Cells. *Biotechnology Bioengineering* 81:430–37.

Nemchinov, L. G., et al. 2000. Development of a Plant-Derived Subunit Vaccine Candidate against Hepatitis C Virus. *Archives of Virology* 145:2557–73.

O'Quinn, P., et al. 2000. Nutritional Value of a Genetically Improved High-Lysine, High-Oil Corn for Young Pigs. *Journal of Animal Science* 78:2144–49.

Pew Initiative on Food and Biotechnology. 2002. Pharming the Field: A Look at the Benefits and Risks of Bioengineering Plants to Produce Pharmaceuticals. http://pewagbiotech.org/.

Potrykus, I. 1999. Vitamin-A and Iron-Enriched Rices May Hold Key to Combating Blindness and Malnutrition: A Biotechnology Advance. *Nature Biotechnology* 17:37.

Richter, L. J., Y. Thanavala, C. J. Arntzen, and H. S. Mason. 2000. Production of Hepatitis B Surface Antigen in Transgenic Plants for Oral Immunization. *Nature Biotechnology* 18:1167–71.

Ruf, S., et al. 2001. Stable Genetic Transformation of Tomato Plastids and Expression of a Foreign Protein in Fruit. *Nature Biotechnology* 19:870–75.

Sijmons, P. C., et al. 1990. Production of Correctly Processed Human Serum Albumin in Transgenic Plants. *Bio/Technology* 8:217–21.

Staub, J., et al. 2000. High-Yield Production of a Human Therapeutic Protein in Tobacco Chloroplasts. *Nature Biotechnology* 18:333–38.

Steinmetz, K. A., and J. D. Potter. 1996. Vegetables, Fruit, and Cancer Prevention: A Review. *Journal of the American Dietetic Association* 96:1027.

Stoll, A., et al. 1999. Omega 3 Fatty Acids in Biopolar Disorder. *Archives of General Psychiatry* 56:407–12.

Stotzky, G. 2000. Persistence and Biological Activity in Soil of Insecticidal Proteins from Bacillus Thuringiensi and of Bacterial DNA Bound on Clays and Humic Acids. *Journal of Environmental Quality* 29:691–705.

Tacket, C. O., et al. 1998. Immunogenicity in Humans of a Recombinant Bacterial-Antigen Delivered in Transgenic Potato. *Nature Medicine* 4:607–9.

————. 2000. Human Immune Responses to a Novel Norwalk Virus Vaccine Delivered in Transgenic Potatoes. *The Journal of Infectious Disease* 182:302–5.

Taylor, S. L., and S. L. Hefle. 2001. Allergic Reactions and Food Intolerances. In *Nutritional Toxicology*, ed. F. N. Kotsonis and M. Mackey. 2nd ed. New York: Taylor & Francis.

Terashima, M. 1999. Production of Functional Human Alpha 1-Antitrypsin by Plant Cell Culture. *Applied Microbiological Biotechnology* 52:516–23.

UNICEF. 1997. A Strategy for Acceleration of Progress in Combating Vitamin A Deficiency. Vitamin A Global Initiative. New York: United Nations.

————. 2000. Vitamin A Deficiency: A Cause of Child Death That Can Be Easily Prevented. Vitamin A Global Initiative. New York: United Nations.

Ursin, V. 2000. Genetic Modification of Oils for Improved Health Benefits. Presentation at conference of the American Heart Association, Reston, Va. June 5–6.

U.S. Department of Agriculture. Animal & Plant Health Inspection Service (APHIS). 2002. Information on field testing of pharmaceutical plants in 2002. http://www.aphis.usda.gov/ppq/biotech/.

————. 2003. USDA Strengthens 2003 Permit Conditions for Field Testing Genetically Engineered Plants. Press release. www.aphis.usda.gov/ppd/rad/ webre por.html.

U.S. Food and Drug Administration and U.S. Department of Agriculture. 2002. Guidance for Industry: Drugs, Biologics, and Medical Devices Derived from Bioengineered Plants for Use in Humans and Animals.

Verwoerd, T. C., et al. 1995. Stable Accumulation of Aspergillus Niger Phytase in Transgenic Tobacco Leaves. *Plant Physiology* 109:1199–1205.

Yu, J., and W. H. Langridge. 2001. A Plant-Based Multicomponent Vaccine Protects Mice from Enteric Diseases. *Nature Biotechnology* 19:548–52.

Zeitlin, L., et al. 1998. A Humanized Monoclonal Antibody Produced in Transgenic Plants for Immunoprotection of the Vagina against Genital Herpes. *Nature Biotechnology* 16:1361–64.

9

Challenging the Misinformation Campaign of Antibiotechnology Environmentalists

Patrick Moore

Environmental thinkers are divided along a sharp fault line. Doomsayers predict the collapse of the global ecosystem; an article featuring my fellow Canadian and famous environmentalist, David Suzuki, talks of "an Apocalypse of Biblical proportions." (Bermingham 1996). Then there are technological optimists who believe that we can feed 12 billion people and solve all our problems with science and technology. They try to debunk all environmental concerns and subscribe to growth for its own sake. Neither of these extremes makes sense. There is a middle road based on science and logic, the combination of which is sometimes referred to as common sense. This is an account of my personal journey of discovery from environmental pessimist to guarded optimist, and why I have come to believe in the promise of agricultural biotechnology.

I was raised in the tiny fishing and logging village of Winter Harbor on the northwest tip of Vancouver Island, where salmon spawned in the streams of the adjoining Pacific rainforest. I didn't realize what a blessed childhood I'd had, playing on the tidal flats by the salmon-spawning streams in the rainforest, until I was shipped away to boarding school in Vancouver at age fourteen. I eventually attended the University of British Columbia, studying the life sciences: biology, forestry, genetics; but it was when I discovered ecology that I realized that through science I could gain an insight into the mystery of the rainforest I had known as a child. I became a born-again ecologist, and in the late 1960s, was soon transformed into a radical environmental activist. A ragtag group of other activists and I sailed a leaky old halibut boat across the North Pacific to block the last hydrogen bomb tests under President Nixon. In the process, I cofounded Greenpeace.

By the mid-1980s, my interest was in "sustainable development," which would take environmental ideas and incorporate them into the traditional social and economic values that govern public policy and our daily behavior. Every morning, 6 billion people wake up with real needs for food, energy, and materials. The challenge is to provide for those needs in ways that reduce negative impact on the environment while also being socially acceptable and technically and economically feasible. Compromise and cooperation among environmentalists, the government, industry, and academia are essential for sustainability.

Not all my former colleagues saw things that way, however. Many environmentalists rejected consensus politics and sustainable development in favor of continued confrontation, ever-increasing extremism, and left-wing politics. At the beginning of the modern environmental movement, Ayn Rand published *Return of the Primitive*, which contained an essay by Peter Schwartz entitled "The Anti-Industrial Revolution." In it, he warned that the new movement's agenda was antiscience, antitechnology, and antihuman (Schwartz 1970, 270). At the time, he didn't get a lot of attention from the mainstream media or the public. Environmentalists were often able to produce arguments that sounded reasonable, while doing good deeds like saving whales and making the air and water cleaner.

But now the chickens have come home to roost. The environmental movement's campaign against biotechnology in general, and genetic engineering in particular, has clearly exposed its intellectual and moral bankruptcy. By adopting a zero-tolerance policy toward a technology with so many potential benefits for humankind and the environment, activists have lived up to Schwartz's predictions. They have alienated themselves from scientists, intellectuals, and internationalists. It seems inevitable that the media and the public will, in time, see the insanity of their position. As my friend Klaus Ammann, Botanical Garden Director at the University of Bern, said to me in January 2001, "Maybe biotech will be the Waterloo for Greenpeace and their allies." Then again, maybe that's just wishful thinking.

Campaign of Fear and Fantasy

On October 15, 2001, I found myself sitting in my office in Vancouver after Greenpeace activists in Paris successfully prevented me from speaking via videoconference to 400 delegates of the European Seed Association. The Greenpeacers chained themselves to the seats in the Cine Cite Bercy auditorium and threatened to shut down the speakers. The venue was hastily shifted elsewhere, but the videoconferencing equipment couldn't be set up at the new location, leading to the cancellation of my keynote presentation.

The issue, in this case, was the application of biotechnology to agriculture, and genetic modification in particular. The conference in Paris was a meeting of delegates from seed companies, biotechnology companies, and government agencies involved in regulation throughout Europe. Surely, these are topics covered by the rules of free speech. Had those rules not been violated, I would have told the assembled that the accusations of "Frankenstein food" and "killer tomatoes" are as much a fantasy as the Hollywood movies from which they are borrowed. I would have argued that, if adding a daffodil gene to rice in order to produce a genetically modified strain can prevent half a million children from going blind each year, then we should move forward carefully to develop it. I would have told them that Greenpeace policy on genetics lacks any respect for logic or science.

In 2001, the European Commission released the results of eighty-one scientific studies on genetically modified organisms, conducted by over 400 research teams at a cost of $65 million. The studies, which covered all areas of concern, did not show

> any new risks to human health or the environment, beyond the usual uncertainties of conventional plant breeding. Indeed, the use of more precise technology and the greater regulatory scrutiny probably make [genetically modified foods] even safer than conventional plants and foods. (European Commission 2001)

Clearly, my former Greenpeace colleagues are either not reading the morning paper, or they simply don't care about the truth. And they choose forcibly to silence those of us who do care about it.

The campaign of fear now waged against genetic modification is based largely on fantasy. In the balance, the real benefits of genetic modification far outweigh the hypothetical and sometimes contrived risks claimed by its detractors.

The programs of genetic research and development now underway in labs and field stations around the world are entirely about benefiting society and the environment. Their purpose is to enhance nutrition, reduce the use of synthetic chemicals, increase the productivity of our farmlands and forests, and improve human health. Those who have adopted a zero-tolerance attitude toward genetic modification threaten to deny these many benefits by playing on fear of the unknown and fear of change.

The case of "Golden Rice" provides a clear illustration of this point. Hundreds of millions of people in Asia and Africa suffer from vitamin A deficiency. Among them, half a million children go blind each year, and millions more suffer from lesser symptoms. Golden Rice has the potential to reduce the suffering greatly, because it contains the gene that makes daffodils yellow, the color of beta carotene, and is the precursor to vitamin A. Dr. Ingo Potrykus, the Swiss co-inventor of Golden Rice, has said that a commercial variety is now ready and available for planting, but also that it will be at least five years before Golden Rice will be able to work its way through the byzantine regulatory system that has been set up as a result of the activists' campaign of misinformation and speculation (Potrykus 2004). So the risk of not allowing farmers in Africa and Asia to grow Golden Rice is that another 2.5 million children will go blind.

What is the risk of allowing this humanitarian intervention to be planted? What possible danger could there be from a daffodil gene in a rice paddy? Yet Greenpeace threatens to rip the genetically modified (GM) rice out if farmers dare to plant it. They have done everything they can to discredit the scientists and the technology, claiming that it would take nine kilos of Golden Rice to alleviate vitamin A deficiency (Greenpeace 2001). Meanwhile Dr. Potrykus has determined that it would only take 100–300 grams of Golden Rice to provide 50 percent of the World Health Organization's recommended daily intake (Beyer and Potrykus 2005). This huge discrepancy can only be accounted for by misinformation on the part of Greenpeace.

The case of Golden Rice is not the only example of civilization being held hostage by activists. Since its introduction to Chinese agriculture in

1996, GM cotton has grown to occupy over one million hectares, or one-third of the total area planted with northern China's most important cash crop. This particular GM variety, called Bt cotton, has been modified to resist the cotton bollworm, its most destructive pest worldwide.

On June 3, 2002, Greenpeace issued a media release announcing the publication of a report on the "adverse environmental impacts of Bt Cotton in China." In typical Greenpeace hyperbole, the public was advised that "farmers growing this crop are now finding themselves engulfed in Bt-resistant superbugs, emerging secondary pests, diminishing natural enemies, destabilized insect ecology," and that they are "forced to continue the use of chemical pesticides" (Ping 2002).

Let's examine these allegations:

- *Bt-resistant superbugs:* There is not a single example or shred of evidence in the Greenpeace report of actual bollworm resistance to Bt cotton in the field. Lab studies in which bollworms were force-fed Bt cotton leaves have yielded some evidence, but any scientist knows that this kind of experiment will eventually result in selection for resistance. Greenpeace, however, is claiming such selection has actually happened to farmers in the field. According to professors Shirong Jia and Yufa Peng of the Chinese National GMO Biosafety Committee, "No resistance of cotton bollworm to Bt has been discovered yet after five years of Bt cotton planting. Resistant insect strains have been obtained in laboratories but not in field conditions" (Jia and Peng 2002). So much for the superbugs.

- *Emerging secondary pests:* Greenpeace points out that there are more aphids, spiders, and other secondary insect pests in fields of Bt cotton than in conventional cotton. This is called an "adverse" impact in their report. The fact is, because Bt cotton requires much less chemical pesticide than conventional cotton, these other insects can survive better in Bt cotton fields. For the scientifically literate, this reduction of impact on nontarget insect species is actually considered one of the environmental benefits of GM crops. How Greenpeace figures this is "adverse" is beyond comprehension.

- *Diminishing natural enemies:* The Greenpeace media release states that there are fewer of the bollworm's natural predators and parasites in Bt cotton fields compared to conventional cotton, and calls this another "adverse impact." Again, a careful reading of the report comes up with no evidence for this claim. And again, according to Professors Jia and Peng, "As of today, there are no adverse impacts reported on natural parasitic enemies in the Bt cotton fields." And, after all, isn't it a bit obvious that if using Bt cotton reduces bollworm populations, bollworm parasite populations will also be reduced? Will Greenpeace now embark on an international campaign to "save the bollworm parasites"?

- *Destabilized insect ecology:* This one is a hoot. To speak of "insect ecology" in a monoculture (where the intention is to have a single species of plant) cotton field that was sprayed with chemicals up to seventeen times a year before the introduction of Bt cotton is ridiculous. The main impact of Bt cotton has been to reduce chemical pesticide use and therefore reduce impacts on nontarget species.

- *Farmers forced to continue using chemical pesticides:* This is the most misleading and dishonest of all the Greenpeace claims. Bt resistance does not give 100 percent protection. Because secondary pests sometimes need to be controlled, farmers using Bt cotton usually use some pesticides during the growing cycle. Professors Jia and Peng sum it up this way:

> The greatest environmental impact of Bt cotton was . . . a significant reduction (70–80 percent) of the chemical pesticide use. It is known that pesticides used in cotton production in China are estimated to be 25 percent of the total amount of pesticides used in all the crops. By using Bt cotton in 2000 in Shandong province alone, the reduction of pesticide use was 1,500 tons. It not only reduced the environmental pollution, but also reduced

the rate of harmful accidents to humans and animals caused by the overuse of pesticides (Jia and Peng 2002).

The Greenpeace report is a classic example of the use of agenda-based "science" to support misinformation and distortion of the truth. Anyone who has studied the introduction of Bt cotton into China and other countries knows that it has resulted in reduced chemical use, reduced impact on nontarget organisms, reduced exposure to chemicals by farm workers, increased productivity, and increased financial benefit to farm owners. To date, there are no substantiated "adverse" impacts from Bt cotton.

GM Crop Technology vs. Greenpeace in the Developing World

Greenpeace consistently demonstrates that its zero-tolerance policy on genetic modification can only be supported by distortions and false interpretations of data—in other words, "junk science." In 2000, Greenpeace launched what turned out to be a twenty-nine-day hunger strike to protest the introduction of Bt corn into the south of the Philippines. In order to whip up media attention, activists spread scare stories that GM corn "would result in millions of dead bodies, sick children, cancer clusters, and deformities" (*Manila Bulletin* 2004). Thankfully, the Philippine government did not give in and stood by its decision, based on three years of consultation and field trials, to allow farmers to plant Bt corn this year. Already there are indications of higher yields and improved incomes for farmers who chose to use the Bt corn.

For six years, activists managed to prevent the introduction of GM crops in India, including Bt cotton. This was largely the work of Vandana Shiva, the Oxford-educated daughter of a wealthy Indian family, who has campaigned relentlessly to "protect" poor farmers from the ravages of multinational seed companies. In 2002, she was given the Hero of the Planet award by *Time* magazine for "defending traditional agricultural practices." *Time* seemed unconcerned by the fact that her defense of these traditional practices in fact needlessly perpetuates poverty and hunger in these communities.

It looked like Shiva would win the GM debate in India until 2001, when unknown persons illegally planted 10,000 hectares of Bt cotton in Gujarat. The cotton bollworm infestation was particularly bad that year, and there was soon a 10,000-hectare plot of beautiful green cotton in a sea of brown. The local authorities were notified and decided that the illegal cotton must be burned. This was too much for the farmers, who could now clearly see the benefits of the Bt variety. In a classic march to city hall, with pitchforks in hand, the farmers protested and eventfully won the day. Bt cotton was approved for planting in March 2002.

Until recently, the situation in Brazil was not promising. A panel of three judges managed to block approval of any GM crops there. Meanwhile, the soybean farmers in the south of the country were quietly smuggling GM soybean seeds across the border from Argentina, where GM soybeans are legal, and at least 8 million tons are produced annually. The fact that Brazil was officially GM-free allowed European countries to import Brazilian soybeans even though they had imposed a moratorium on the importation of GM crops. All that Europe really required was a wink and a nudge, and the ability to pretend that Brazil didn't grow any GM soybeans.

With the election of Brazilian president "Lula" da Silva of the Workers' Party in 2002, however, the "green" elements within the party pressed the government to enforce the ban on genetically modified organisms. There was something ironic about a Workers Party enforcing a policy that would damage farmers who had come to enjoy the benefits of GM technology. In the end, the Brazilian farmers rebelled like those in India. In 2003, the government relented and allowed GM soybeans to be planted. The soybean farmers of southern Brazil have since become prosperous, bringing benefits to the environment and their local communities.

Breaking Greenpeace's Grip on the Media

Surely there is some way to break through the misinformation and hysteria and get a more balanced picture to the public. Surely if reasonable people who support Greenpeace saw the choice between the risk of a daffodil gene in a rice plant versus the certainty of millions of blind children, they would descend on the organization's offices around the world and demand to have their money back. How is it that these charlatans continue to stymie progress on so many fronts when their arguments are nothing more than wild, scary speculation?

The main reason for the failure to win the debate decisively is the failure of supporters of GM technology to act decisively. The activists are playing hardball while the GM side is soft-pedaling the health and environmental benefits of this new technology. Biotech companies and their associations use soft images and calm language, apparently to lull the public into making pleasant associations with GM products. How can that strategy possibly hope to counter the Frankenfood fears and superweed scares drummed up by Greenpeace and others?

A brief scan of the Monsanto, Syngenta, and Council for Biotechnology Web sites shows clearly that they are trying to project positive, clean, and calming thoughts (Monsanto 2005; Syngenta 2005; Council for Biotechnology 2005). This is all well and good, but it is no way to turn the tide. Stronger medicine must be prescribed. Imagine an advertising campaign that showed graphic images of blind children in Africa, explained vitamin A deficiency, introduced Golden Rice, and described how Greenpeace's actions are preventing the delivery of this cure. Imagine another ad that showed impoverished Indian cotton farmers, explained Bt cotton, and gave the statistics for increased yield, reduced pesticide use, and betterment for farmers. Again, it would be clearly stated that activists were the reason for the delayed adoption of the technology. How about an ad that graphically portrayed the soil erosion and stream siltation caused by conventional farming, versus the soil conservation obtained by using GM soybeans and canola? And another one that showed workers applying pesticides without protection in a developing country, versus the greatly reduced applications possible with Bt corn and cotton? What if all these ads were hosted by a well-known and trusted personality? Wouldn't this change public perspectives?

The biotechnology sector needs to ramp up its communications program, and to get a lot more aggressive in explaining the issues to the public through the media. Nothing less will turn the tide in the battle for the minds, and hearts, of people around the world.

References

Bermingham, John. 1996. It's Just Too Hot for Us to Handle: Disaster Looms, Warns Scientist Suzuki. *Province* (Vancouver, B.C.). January 5. A5.

Beyer, Peter, and Ingo Potrykus. 2005. How Much Vitamin A Rice Must One Eat? *AgBioWorld.* http://www.agbioworld.org/biotech-info/topics/goldenrice/how_much.html (accessed April 25, 2005).

Council for Biotechnology. 2005. Council for Biotechnology Web site. http://www.whybiotech.com/.

European Commission. 2001. *A Review of Results: EC-Sponsored Research on Safety of Genetically Modified Organisms*, ed. Charles Kessler and Loannis Economidis. Luxembourg: Office for Official Publications of the European Communities, L-2985. October 9.

Goklany, Indur M. 2000. Applying the Precautionary Principle to Genetically Modified Crops. Policy Study No. 157. Center for the Study of American Business. August.

Greenpeace. 2001. Genetically Engineered "Golden Rice" Is Fool's Gold. Media Release. Manila/Amsterdam. February 9.

Heath, Clark W., Jr. 1997. Pesticides and Cancer Risk. *Cancer* 80:1887–88.

Jia, Shirong, and Yufa Peng. 2002. Chinese Scientists Criticize Greenpeace Bt Report. Media Release. August.

Manila Bulletin. 2004. Local Savants Dare GMO Activist to Show Conclusive Evidence. March 15.

Monsanto. 2005. Monsanto Web site. http://www.monsanto.com/monsanto/layout/default.asp.

Ping, Lo Sze. 2002. Chinese Experience Shows Adverse Environmental Impacts of Genetically Engineered BT Cotton. Greenpeace Media Release. Beijing/London. June 3.

Potrykus, Ingo. 2004. Experience from the Humanitarian Golden Rice Project: Extreme Precautionary Regulation Prevents Use of Green Biotechnology in Public Projects. *BioVision.* Alexandria. April.

Schwartz, Peter. 1970. The Anti-Industrial Revolution. In Ayn Rand, *Return of the Primitive.* New York: Meridian.

Syngenta. 2005. Syngenta Web site. http://www.syngenta.com/en/index.aspx.

Postscript

The only winner in the current stalemate over genetically modified foods is the protest industry, particularly self-designated "green" groups that call for public debate but already have their minds closed to the science. We are just beginning to exploit the potential of genetic technology. There is the promise of lifesaving drugs tailored to our individual genetics, the eventual disappearance of congenital diseases, and the remarkable manipulation of cells to regrow bones and damaged organs, including our hearts and brains. However, these scientific advances are years, if not decades, in the future. Only in agriculture is the promise of biotechnology actually being realized. Let's hope that reasoned discourse and a careful balancing of risk against lost opportunity can help restore public confidence so vital research can continue.

We need to go beyond the precautionary principle and its obsession with "worst case" risks to a "risk-risk" model that also takes into account the potential benefits to change, not just potential dire consequences. While the precautionary principle states that the overriding goal of public policy should be guided by the ultraconservative drive to eliminate risk, a risk-risk approach would evaluate a situation based on the trade-offs that inevitably occur when reasonable risks are avoided at all cost. GM technological innovations should be evaluated not just on the basis of the farfetched dangers they potentially pose, but on the often-horrific, day-to-day harm that not risking innovation imposes on the most vulnerable members of society. How many children in the developing world have to die before we can risk moving forward with this technology?

About the Authors

Jay Byrne is president of v-Fluence Interactive Public Relations. Mr. Byrne has twenty years of experience in public relations, campaign communications, and government affairs, formerly holding communications positions at the White House, the U.S. State Department, and the Monsanto Company, and for the City of Boston. v-Fluence is a public affairs, issues-management, and marketing support firm with special expertise in biotechnology and online reputation management.

Gregory Conko is a senior fellow and director of food safety policy with the Competitive Enterprise Institute, a public interest group based in Washington, D.C., where he specializes in issues of food and pharmaceutical drug safety regulation, and the general treatment of health risks in public policy. Mr. Conko is also vice president and a member of the board of directors of the AgBioWorld Foundation, which he cofounded. AgBioWorld Foundation provides information to teachers, journalists, policymakers, and the general public about developments in plant science, biotechnology, and sustainable agriculture.

Jon Entine is a scholar in residence at Miami University (Ohio) and an adjunct fellow at the American Enterprise Institute. He has written *Taboo: Why Black Athletes Dominate Sports and Why We're Afraid to Talk about It* (PublicAffairs 2000) and writes frequently in popular and academic publications on science, public policy, and business ethics.

Carol Tucker Foreman is the director of the Consumer Federation of America's Food Policy Institute, and the former assistant secretary of agriculture for food and consumer services. She is also a member of the EU/US Consultative Forum on Biotechnology and the Agricultural Policy Advisory Committee for Trade. Ms. Foreman served on the U.S. Department of Agriculture's Advisory Committee on Agricultural Biotechnology and oversaw the development of the federal government's first *Dietary Guidelines for Americans.*

Tony Gilland is the science and society director of the British Institute of Ideas. Mr. Gilland contributes to a range of publications, radio programs, and public events addressing science controversies, including the Institute of Ideas' "Genes and Society Festival" in London in 2003 and the institute's and Royal Institution's "Interrogating the Precautionary Principle" conference in 2000. He also edited *Science—Can We Trust the Experts?* (Hodder & Stoughton, 2002) and *Nature's Revenge?* (Hodder & Stoughton, 2002).

Thomas Jefferson Hoban is a professor in the department of sociology and anthropology and the department of food science, and the director of the Center for Biotechnology in a Global Society at North Carolina State University. Mr. Hoban is also a member of the U.S. Department of Agriculture's Advisory Committee on Agricultural Biotechnology and a former member of the U.S. Food and Drug Administration's biotechnology labeling panel.

Patrick Moore is a founding member of Greenpeace, former president of Greenpeace Canada, and former director of Greenpeace International. As the leader of many environmental campaigns, Mr. Moore was a driving force shaping policy and direction as Greenpeace became the world's largest environmental activist organization. He is currently head of the environmental group Greenspirit, Vancouver, B.C., which he founded, and which focuses on environmental policy and communications in natural resources, biodiversity, energy, and climate change. He previously was a member of British Columbia's government-appointed Round Table on the Environment and Economy from 1990 to 1994.

Andrew S. Natsios is the administrator of the U.S. Agency for International Development. He served previously at USAID, first as director of the Office of Foreign Disaster Assistance from 1989 to 1991, and then as assistant administrator for the Bureau for Food and Humanitarian Assistance (now the Bureau of Democracy, Conflict and Humanitarian Assistance) from 1991 to 1993. Previously, Mr. Natsios headed the Massachusetts Turnpike Authority and served as secretary for administration and finance for the Commonwealth of Massachusetts, vice president of World Vision U.S., and executive director of the Northeast Public Power Association.

Martina Newell-McGloughlin is the director of the University of California Systemwide Biotechnology Research and Education Program, which is based at the University of California–Davis. Ms. Newell-McGloughlin is also the codirector of the NIH Training Program in Biomolecular Technology, and a member of the World Trade Organization's Genomics Panel on Technology, the International Food Information Council Expert Panel, and the United Nations Technology Discussion Panel on Sustainable Agriculture.

Robert L. Paarlberg is a professor of political science at Wellesley College and an associate of the Harvard University Center for International Affairs. He is also a consultant with the International Food Policy Research Institute, USAID, the U.S. Department of Agriculture, and the U.S. State Department. Mr. Paarlberg's books include *The Politics of Precaution: Genetically Modified Crops in Developing Countries* (International Food Policy Research Institute, 2001) and *Leadership Abroad Begins at Home: U. S. Foreign Economic Policy after the Cold War* (Brookings Institution Press, 1995).

C. S. Prakash is a professor of plant biotechnology at Tuskegee University and president of the AgBioWorld Foundation based in Auburn, Alabama, which operates an internationally respected Web site, www.agbioworld.org, which disseminates information and promotes discussion on this subject among food biotechnology stakeholders, such as scientists, policymakers, activists, and journalists. He recently served on the U.S. Department of Agriculture's Agricultural Biotechnology Advisory Committee and continues to serve on the Advisory Committee for the Department of Biotechnology for the government of India.

Index

Murray L. Weidenbaum
Mallinckrodt Distinguished
University Professor
Washington University

Richard J. Zeckhauser
Frank Plumpton Ramsey Professor
of Political Economy
Kennedy School of Government
Harvard University

Research Staff

Joseph Antos
Wilson H. Taylor Scholar in Health
Care and Retirement Policy

Leon Aron
Resident Scholar

Claude E. Barfield
Resident Scholar; Director, Science
and Technology Policy Studies

Roger Bate
Resident Fellow

Walter Berns
Resident Scholar

Douglas J. Besharov
Joseph J. and Violet Jacobs
Scholar in Social Welfare Studies

Edward Blum
Visiting Fellow

Dan Blumenthal
Resident Fellow

Karlyn H. Bowman
Resident Fellow

John E. Calfee
Resident Scholar

Charles W. Calomiris
Visiting Scholar

Lynne V. Cheney
Senior Fellow

Steven J. Davis
Visiting Scholar

Veronique de Rugy
Research Fellow

Thomas Donnelly
Resident Fellow

Nicholas Eberstadt
Henry Wendt Scholar in Political
Economy

Mark Falcoff
Resident Scholar Emeritus

Gerald R. Ford
Distinguished Fellow

John C. Fortier
Research Fellow

Ted Frank
Resident Fellow; Director,
AEI Liability Project

David Frum
Resident Fellow

Ted Gayer
Visiting Scholar

Reuel Marc Gerecht
Resident Fellow

Newt Gingrich
Senior Fellow

James K. Glassman
Resident Fellow

Jack L. Goldsmith
Visiting Scholar

Robert A. Goldwin
Resident Scholar Emeritus

Michael S. Greve
John G. Searle Scholar

Robert W. Hahn
Resident Scholar; Director,
AEI-Brookings Joint Center
for Regulatory Studies

Kevin A. Hassett
Resident Scholar; Director,
Economic Policy Studies

Steven F. Hayward
F. K. Weyerhaeuser Fellow

Robert B. Helms
Resident Scholar; Director,
Health Policy Studies

Frederick M. Hess
Resident Scholar; Director,
Education Policy Studies

R. Glenn Hubbard
Visiting Scholar

Frederick W. Kagan
Resident Scholar

Leon R. Kass
Hertog Fellow

Jeane J. Kirkpatrick
Senior Fellow

Herbert G. Klein
National Fellow

Marvin H. Kosters
Resident Scholar Emeritus

Irving Kristol
Senior Fellow Emeritus

Desmond Lachman
Resident Fellow

Michael A. Ledeen
Freedom Scholar

Adam Lerrick
Visiting Scholar

James R. Lilley
Senior Fellow

Lawrence B. Lindsey
Visiting Scholar

John R. Lott Jr.
Resident Scholar

John H. Makin
Visiting Scholar

N. Gregory Mankiw
Visiting Scholar

Allan H. Meltzer
Visiting Scholar

Joshua Muravchik
Resident Scholar

Charles Murray
W. H. Brady Scholar

Roger F. Noriega
Visiting Fellow

Michael Novak
George Frederick Jewett Scholar
in Religion, Philosophy, and Public
Policy; Director, Social and Political
Studies

Norman J. Ornstein
Resident Scholar

Richard Perle
Resident Fellow

Alex J. Pollock
Resident Fellow

Sarath Rajapatirana
Visiting Fellow

Michael Rubin
Resident Scholar

Sally Satel
Resident Scholar

Gary Schmitt
Resident Scholar

Joel Schwartz
Visiting Fellow

Vance Serchuk
Research Fellow

Daniel Shaviro
Visiting Scholar

Christina Hoff Sommers
Resident Scholar

Phillip L. Swagel
Resident Scholar

Samuel Thernstrom
Managing Editor, AEI Press;
Director, W. H. Brady Program

Fred Thompson
Visiting Fellow

Peter J. Wallison
Resident Fellow

Scott Wallsten
Resident Scholar

Ben J. Wattenberg
Senior Fellow

John Yoo
Visiting Scholar

Karl Zinsmeister
J. B. Fuqua Fellow; Editor,
The American Enterprise